CELL AND TISSUE
REGENERATION

CELL BIOLOGY: A SERIES OF MONOGRAPHS

E. Edward Bittar, Series Editor

CELL AND TISSUE REGENERATION

A Biochemical Approach

MARGERY G. ORD
LLOYD A. STOCKEN
University of Oxford

A Wiley-Interscience Publication

JOHN WILEY & SONS

New York • Chichester • Brisbane • Toronto • Singapore

Library of Congress Cataloging in Publication Data:

Ord, Margery G.
  Cell and tissue regeneration.

  (Cell biology: a series of monographs; v. 2)
  "A Wiley-Interscience publication."
  Bibliography: p.
  Includes index.
  1. Regeneration (Biology) 2. Cytochemistry.
3. Histochemistry. I. Stocken, Lloyd A. II. Title.
III. Series. [DNLM: 1. Regeneration. 2. Cells—
Physiology. QH 499 065c]
QH499.073 1984   591.8   84-3536
ISBN 0-471-86248-7

Printed in the United States of America

10 9 8 7 6 5 4 3 2 1

*To Noreen
and our colleagues*

SERIES PREFACE

The aim of the Cell Biology Series is to focus attention upon basic problems and show that cell biology as a discipline is gradually maturing. In its largest aim, each monograph seeks to be readable and informative, scholarly, and the work of a single mind. In general, the topics chosen deal with major contemporary issues. Together they represent a rather large domain whose importance has grown enormously in the course of the last generation. The introduction of new techniques has no doubt ushered in a small revolution in cell biology. However, we still know very little about the cell as an ordered structure. As will become abundantly clear to the reader, real progress is not just a matter of progress of technique but also a matter of close interaction between advances in different fields of study, as well as genesis of new approaches and generalized concepts.

E. EDWARD BITTAR

Madison, Wisconsin
January 1984

PREFACE

The ability or inability of cells, tissues, and organs to regenerate is a topic which has fascinated biologists ever since it was noted that worms can repair themselves after truncation, newts regrow their tails, and humans, who can compensate physiologically for many tissue injuries, have no capacity to restore their limbs. Some thorough and authoritative reviews of regeneration have been written within the last 20 years, but until recently it was not possible to attempt a biochemical interpretation of the results from the sophisticated experimental manipulations now being attempted with *Hydra*, *Planaria*, or urodele limbs. We have had to make extensive use of biochemical data from cell cultures and from mammalian tissues showing compensatory hyperplasia. Obviously there are dangers in the use of analogy, but we believe analytical methods now being applied in mammalian systems are so sensitive and powerful that it may be valuable to consider results from current molecular biology that may be applicable to growth in adult tissues and to the potential for regeneration.

We have not attempted to provide a definitive account of all invertebrate and vertebrate systems where regeneration occurs. We have selected well-studied systems and used examples of compensatory hyperplasia in mammals for which extensive biochemical data are available. Graduates in biochemistry, genetics, or molecular biology are often unaware of problems in cell biology, whose complexity and open-endedness they find distracting, especially in the context of final examinations. Conversely, biology students may not realize how much the work in areas more easily interpreted molecularly has provided results which may be of great relevance in their more complex fields.

This book falls into three parts: the systems chosen to illustrate regeneration and compensatory hyperplasia; the signals eliciting growth, proliferation, and tissue restructuring; and the responses shown by the cell membrane, its cytoplasm, and its nucleus. Where possible, review references are given to cover the literature before 1981; we regret the consequent exclusion of many important individual contributions. Most of the material for inclusion was collected by the end of June, 1983.

We hope this volume may attract the attention of final year and graduate biochemists and biologists to regeneration, whose study is now providing results relevant to the outstanding problem in biology, namely, growth and differentiation.

We could not have attempted this book without a great deal of help from our colleagues in this and other departments of the university. We are also grateful to many other friends at home and abroad, who sent us their reprints, and to the organizers of the Harden Conference in September, 1982, "Cell Cycles," and the European Molecular Biology Organisation workshop in April, 1983, "Growth Control in Normal and Malignant Cells," whose programs were very useful to us.

We are especially grateful to Dr. Pat Fitzgerald, Professor Henry Harris, Dr. Jane Karlsson, Dr. John McLachlan, Dr. David Roberts, Mr. Paul Schofield, and Dr. Tony Watts for reading the manuscript. The mistakes are still ours, but their help was great and much appreciated.

We also wish to thank Frank Caddick for his cooperation and skill in preparing the figures; Louise Sear struggled bravely and successfully to process the manuscript; and Brian Taylor was very helpful with the references.

MARGERY G. ORD
LLOYD A. STOCKEN

Oxford, England
April 1984

CONTENTS

ABBREVIATIONS

A 23187	Ca ionophore
ACTH	adrenocorticotropin
ADP-ribose	adenosine diphosphate ribose
AER	apical ectodermal ridge
BFU	burst-forming unit
bp, kbp	base pair (in DNA sequence), kilo–base pair
cDNA	DNA copy of mRNA
CFU	colony-forming unit
CM	conditioned medium
conA	concanavalin A
DMSO	dimethyl sulfoxide
EGF	epidermal growth factor
eIF	eukaryotic initiation factor (for protein synthesis)
eIF-2P	phosphorylated eIF-2
EM	electron microscope
F(D)GF	fibroblast (derived) growth factor
Gy	gray (unit of radiation, $= 100$ rads)
Hb	hemoglobin
HIM	hemopoietic inductive microenvironment

HMBA	hexamethylene bisacetamide
HMG	highly mobile group of nonhistone chromosomal proteins
HMG-CoA	3-hydroxy, 3-methyl glutaryl coenzyme-A
HnRNA	heterogeneous nuclear RNA
HnRNP	heterogeneous ribonucleoprotein (particle)
IGF	insulinlike growth factor
K_D	dissociation constant
kDa	kilodalton (unit of molecular mass)
K_m	Michaelis constant
LDL	low-density lipoprotein
^mC	5-methyl cytosine
M_r	molecular mass relative to $O = 16$
met tRNA$_f$	methionyl tRNA which can be formulated (initiating aminoacyl-tRNA for protein synthesis)
MSA	multiplication-stimulating activity
MSH	melanophore-stimulating hormone
NAD	nicotinamide adenine dinucleotide
NHCP	nonhistone chromosomal protein
P_i	inorganic phosphate
PDGF	platelet derived growth factor
PHA	phytohemagglutinin
poly(A)$^+$mRNA	3′polyadenylated mRNA
RBC	red blood cells
RNA pol I, II, III	RNA polymerases; I transcribes rRNA, II mRNA, and III, 5s and tRNAs
SDS	sodium dodecylsulfate
TdR	thymidine
ZPA	zone of polarizing activity

COMMON CELL LINES

E	epithelial origin
F	fibroblast origin

BHK	baby Syrian hamster kidney-E
BSC-1	African green monkey-E
CHO	Chinese hamster ovary-E
C11D	mouse cell line
HeLa	human cervical carcinoma-E
HTC	rat Morris hepatoma-F
L	mouse cell line-F
P815	mouse cell line
3T3	Swiss mouse embryo-F

GENETIC AND VIRAL ABBREVIATIONS

Amy-1A	amylase gene
ars	autonomously replicating sequence
cdc	cell division cycle mutant
en	engrailed (*Drosophila* homoeotic mutant)
ltr	long terminal repeat (DNA sequence)
mwh	multiple wing hairs (*Drosophila* mutant)
onc	oncogene
rsv	Rous sarcoma virus
src	oncogene from Rous sarcoma virus
SV4O	simian virus
tk	thymidine kinase (gene)
ts	temperature sensitive (conditional mutant)

1

THE SYSTEMS—CELLS, CELL CULTURES, AND INVERTEBRATES

In the animal kingdom the capacity for regeneration is most evident in the invertebrates; in vertebrates regeneration is very limited, although cellular proliferation and restitution of function are shown by some tissues in all classes of vertebrates. In Amphibia salamanders and newts regenerate ablated limbs or tails, but adult frogs do not normally regenerate limbs, and the capacity for regeneration in fish is very restricted. In the plant kingdom the capacity to regenerate is widespread and may be universal.

Our discussions will start with eukaryotic cell cultures since analysis of the factors determining cell growth and proliferation is fundamental to our understanding of growth generally and the processes which regulate it.

1.1.   SIMPLE GROWTH CYCLES

Experimental data on which to base analyses of growth curves have come mainly from bacteria, yeast, and other eukaryotic cell cultures. Various approaches are available. First, there is population kinetics, where statistical analyses are made of the increase in the number of cells over time. Second, especially with the fission yeast *Schizosaccharomyces pombe* and budding yeast *Saccharomyces cerevisiae*, measurements can be made on individual cells by time lapse cinematography (Wheals, 1982); with other cell types, flow microfluorometry can be used to determine the size of the cell and its DNA content (Yen & Pardee, 1979). These methods, which have been developed during the last 10 years, have focused attention on differences in behavior between daughter cells and thus form the basis of probabilistic models of cell cycles.

To follow molecular changes through the cell cycle, synchronized populations are commonly needed to provide sufficient material for biochemical analysis. For some studies of macromolecules, labelling with very high specific radioactivity precursors, immunofluorescent techniques, and sensitive gel electrophoretic methods allow analyses to be performed with a very small number of cells (even one). Usually such methods are unsuitable for lower-molecular-weight compounds.

Three types of synchrony can be distinguished (Mitchison, 1971; Prescott, 1976): natural, selected, and induced. Natural synchrony is seen in early stages in embryonic development; the first cleavage cycle in fertilized sea urchin eggs is a well-studied example. Nuclear division in the acellular slime mold *Physarum polycephalum* is another favorite, as are also germinating spores.

Selection of synchronized cells is based on differences in their physical properties through the cycle. Mitotic shaking was introduced by Terasima and Tolmach (1961) to separate rounded-up, minimally adhesive HeLa (human cervical carcinoma-E) cells in mitosis. Synchrony by this means is excellent, but the number of cells which can be obtained is limiting, mitosis being a rather small part (c. 5%) of the cell cycle. Procedures to accumulate the detached cells tend to blur the synchrony and produce other disturbances. For some cell types separation by differences in size and density through the cycle can be achieved by zonal or elutriating centrifugation. Relatively good yields with minimal side effects can be obtained, but the distinction between the different stages may not be very sharp, and the process of centrifugation can produce detectable perturbations in the properties of the cells. (For critical discussion of methodology see Lloyd et al., 1982).

Synchrony can be induced in a number of ways: alterations in the environmental temperature (Tetrahymena; Scherbaum & Zeuthen, 1954), light (Chlorella; Tamiya, 1963), nutrient availability (serum deprivation or the selective deprivation of essential amino acids such as isoleucine), or the use of reversible inhibitors. Thymidine, which probably through its conversion to TTP (thymidine 5'triphosphate) and feedback inhibition of ribonucleotide reductase and thus DNA synthesis (Reichard, 1972; Hunting & Henderson, 1981), blocks progress through the cycle at the G_1/S transition, and colcemid related derivatives, which arrest cells in mitosis by affecting microtubules in the spindle apparatus, are probably the best-known inhibitors. Induction methods are very commonly and easily used and can be applied on a relatively large scale. Unfortunately, the methods are notorious for their undesirable and sometimes unappreciated secondary effects, which can cause unexpected and misleading changes in cell behavior.

1.1.1. Bacterial Cell Cultures (Mandelstam et al., 1982)

Two features emerge from studies on bacterial cell growth which have profoundly affected thinking about eukaryotic cell cycles: the extreme variability in mean generation times shown by populations of any given microorganism growing at a fixed temperature but in different culture media, and the major influence cell size has on cell cycle times. With *E. coli* at 37°C on a simple glucose-containing medium, the mean generation time is about 60 min. If a peptide-enriched medium is used, so providing a plentiful supply of amino acids, the mean generation time drops to about 20 min; response to changes in the culture medium occurs very quickly. With conditions which support less-rapid growth, up to eightfold variation in cell volume may be found. When this occurs, however, it is observed that all cells have the same volume 60 min before DNA synthesis is initiated and, further, that all cells have the same length 20 min before septation is initiated, leading to cell division (Donachie, 1981).

Cells which are growing rapidly have more ribosomes; cell size depends on the number of ribosomes the cell contains and large cells have shorter generation times. In a population which has reached stationary phase the cells are smaller and more slowly growing than those in log-phase cultures.

In bacteria, DNA is being replicated continuously through the cell cycle, and cells with rapid growth rates have multiple sites on their DNA from which replication is proceeding. Nuclear division cycles and cell division cycles are thus not tightly coupled but may appear to be linked through the operation of the SOS system (Little & Mount, 1982), whereby any interference in DNA replication causes cell division to be switched off.

The effects which culture conditions, notably the energy source and the availability of amino acids, have on the number of ribosomes/cell, cell size, and growth rate lead to a deterministic view of events in the cell cycle, but with considerable ignorance over the sizing mechanism. This could be positive or negative, responding to an increase in concentration of some critical factor or the the dilution of an inhibitor, with the sensing mechanism chromosomal or in the cytoplasm (Fantes & Nurse, 1981).

1.1.2. Eukaryotic Cell Cultures

SIMPLE GROWTH CYCLES (FIG. 1.1)

DNA synthesis is a periodic function in eukaryotic cells. In most but not all cells it occurs in interphase, separated from the end of mitosis by a period in G_1 and followed by a usually shorter gap, G_2, before the next mitosis starts. This sequence was first demonstrated by Howard and Pelc (1951) with $^{32}P_i$ labelling experiments in *Vicia faba*. In many cell types the times for DNA synthesis (S phase) (6–8 h), G_2 (6–8 h), and mitosis (1 h) are almost invariant, but considerable variations occur in the length of G_1. Further, if individual cells are examined, transit times through G_1 may show coefficients of variation of up to 20% between the two daughter cells, introducing a probabilistic factor into the analysis. With the budding yeast *Saccharomyces cerevisiae*, in which the daughter cell is normally smaller than its mother, its cycle time is always the longer (see Wheals, 1982). As

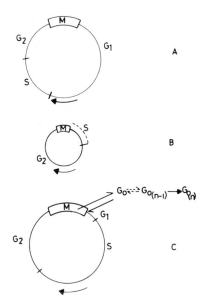

Figure 1.1. Cell Cycles. (*a*) The classical cycle (after Howard & Pelc, 1951). Approximate duration in mammalian cells, 24 h. (*b*) Cleavage cycle in embryos. (*c*) Archetypal cell cycle (after D. M. Prescott, personal communication).

progressive changes in mean mass or size/cell do not occur, growth and division must be coordinated, emphasizing, as with bacterial cultures, the involvement of a size-dependent determinant also.

Although the behavior of many cells in culture or in the whole animal is consistent with their progress through a cycle of the classical type (see Fig. 1.1A), exceptions to this are well established (see Prescott, 1982). Most obvious for normal cells are those in cleavage-stage embryos where DNA synthesis commences in mitosis, usually in telophase. There is no G_1 phase (see Fig. 1.1B). CHO (Chinese hamster ovary-E) cell lines are also known which show neither G_1 nor G_2 phases (Rao et al., 1978; Prescott, 1982). These exceptions must be considered when cell cycle controls are reviewed (see Restriction Points, following). The most extreme modifications from the classical cycle suggest that, in archetypal cells, DNA synthesis commences as soon as metaphase is complete and that, in these cells also, the factors necessary for chromosome separation are not lost once mitosis is ended, but are immediately available when the 4n amount of DNA has been replicated. More relevant to regeneration of adult somatic tissues, it is possible that cells destined to move out of the growth cycle can exist in a number of states with altered phenotype (see Fig. 1.1C); changes (Chapters 6–8) associated with movement from the G_0 to the G_1 state could easily be considered as a progression from a more to a less quiescent state. A restriction point in G_1 (see following) would then mark a critical transition in this progress out of quiescence.

RESTRICTION POINTS (PARDEE ET AL., 1978)

A restriction point in a cell cycle is a term associated especially with Pardee and his colleagues, marking a critical transition in the progress of the cell after which it becomes committed to a different functional state. Originally restriction points were detected in G_1, after which the cell commenced DNA replication. Many different lines of evidence support the existence of this critical transition in mammalian cell cultures. The presence of essential growth factors in the medium is required early in G_1 but can be withdrawn when the critical point is traversed. Low concentrations of cycloheximide (Rossow et al., 1979) given early in G_1 markedly extend the time spent in this phase, but if they are given in late G_1, transit through the end of G_1, S, G_2, and M are hardly affected. Low concentrations of actinomycin can be used to give results similar to those with cycloheximide. The restriction point so identified may be about 2 h before S phase in animal cells commences (Pardee et al., 1978). Other stages in the cycle can show similar properties, for example, the transitions from G_2 into M where specific but unidentified RNA and protein synthesis are again required (see Rao, 1980).

Results from these and other (see Cell Division Cycle Mutants, following) approaches suggest the presence of a protein which is essential for transition through a restriction point. Kinetic characteristics for such proteins have been postulated (see Campisi et al., 1982), but their identification and the reconciliation of data from types of experiments described above with that from genetic investigations has not yet been achieved.

With some types of cell cycle (see Fig. 1.1B), notably the first cleavage cycle in sea urchins, protein synthesis is not required. Activation of the cell membrane, not necessitating the presence of a sperm, initiates a series of membrane changes leading to replication, mitosis, and cytokinesis. Critical factors for transit through restriction points are phenotypic characteristics of cells.

CELL DIVISION CYCLE (cdc) MUTANTS

One promising approach to the identification of regulatory factors required before traversing restriction points is the development of mutants which have blocks at identifiable stages in the cell cycle. These mutants are necessarily conditional and are usually temperature sensitive (ts), being unable to progress through the cycle at the nonpermissive temperature. Frequently the mutants produce easily observable cytological or functional defects. These then have to be related to the alteration in, or absence of, the gene product. A further difficulty for cell cycle analysis is that the time at which the gene is transcribed need not be the time when the gene product is essential (Hartwell, 1978; Pringle & Hartwell, 1981).

Much effort is devoted to *Saccharomyces cerevisiae* since yeasts grow and divide in haploid as well as diploid states, thus greatly assisting genetic analysis. Well over 100 cdc mutants are now identified in *S. cerevisiae*, affecting, for example, polar body duplication or separation, initiation of DNA synthesis, cytokinesis, and bud emergence. A stage analogous to the restriction point in G_1 in mammalian cell cultures can be recognized— *start*—after which cells are committed to the mitotic cycle (Hartwell, 1978). The step obligatorily requires the cdc-28 gene product (see Carter, 1981; Yanishevsky & Stein, 1981). Similar genetic approaches are in hand for fission yeast (Nurse & Fantes, 1981), through the production of *wee* mutants which divide when smaller than wild-type cells, and for *Aspergillus* and *Tetrahymena*, as well as for numerous mammalian cell types. It is encouraging for the universalist view of cell cycle events that complementation has now been shown between cdc-28 gene product(s) in *S. cerevisiae* and those of cdc (wee-2) in *S. pombe* (Nurse & Fantes, 1981; Beach et al., 1982). The cdc-28/2 gene product is also required when cells pass from G_2 phase into mitosis; their gene sequences are not homologous (Beach et al., 1982).

Nuclear-Cytoplasmic Interactions in Cell Cycle Regulation

Nuclear involvement in the program for sequential synthesis of proteins required for successive stages in the cell cycle is complementary to the involvement of the cytoplasm in switching this program on and off. Evidence for the importance of the cytoplasm came from nuclear transplantation experiments in the mid 1950s (see Gurdon, 1974) and from cell fusion experiments in the 1960s (see Harris, 1970; Rao & Johnson, 1970).

In the former, nuclei from adult frog brain were injected into the cytoplasm of three different recipients: ovarian oocytes which were synthesizing RNA, especially rRNA; ovulated oocytes which were completing meiosis and had condensed chromosomes; and unfertilized eggs, where RNA synthesis had stopped. After injection, the transplanted nuclei respectively commenced

RNA synthesis, showed chromosome condensation, or initiated DNA synthesis, that is, nuclei of identical origin showed a pattern of behavior determined by the recipient cytoplasm (Johnson & Rao, 1970).

Essentially similar results were obtained following the formation of HeLa-chick heterokaryons (Johnson & Harris, 1969). By use of the Sendai virus technique chick erythrocytes were fused with human HeLa cells. The "switched-off" erythrocyte nuclei swelled, their chromatin redispersed, and DNA synthesis started in synchrony with that in the host HeLa nucleus. Besides showing the influence HeLa cytoplasm and/or proteins programmed from the HeLa nucleus had on expression by the donor nucleus, the experiments also demonstrated strikingly the absence of species specificity in the factors involved.

With HeLa homokaryons (Rao & Johnson, 1970) fusions could be achieved between cells in different phases in the cell cycle. G_1/S fusions allowed DNA synthesis in the G_1 nucleus to be induced, but if G_2/S fusion occurred, DNA synthesis in the S phase nucleus was not affected by the presence of the G_2 nucleus, nor was DNA synthesis in that nucleus affected by the nucleus in S. Dosage effects were seen, DNA synthesis being more evident in G_1 nuclei in $G_1/2S$ and $G_1/3S$ homokaryons than in $G_1/1S$. Nuclei which entered mitosis detectably ahead of other nuclei in the homokaryon produced premature chromosome condensation in the other nuclei (Johnson & Rao, 1970). Refinements in this type of experiment (see Rao, 1980) have confirmed that in heterokaryons nonspecific cytoplasmic factors promote nuclear DNA synthesis but that the time required for nuclei to go through S phase usually remains characteristic of the original cell type. It is generally concluded that S phase cytoplasm contains factors positively stimulating nuclei to go into S phase rather than that an inhibitor present at the start of G_1 becomes progressively diluted. The alternative interpretation, of an endogenous inhibitor of cell proliferation being present in resting (3T3—Swiss mouse embryo-F) cells, cannot be precluded (Polunovsky et al., 1983).

The results of heterophasic fusion can also be studied in *Physarum* (Sachsenmaier, 1981), which has no G_1 phase. When equal areas of plasmodia were taken, older nuclei, which were far into G_2 at the time of fusion, were delayed while younger nuclei caught up. The temporal advance of the younger nuclei was always shorter than the temporal delay of the older nuclei. Protein:DNA ratios at mitosis were constant. Evidence of a critical point in G_2 was provided; nuclei within 20 min from the onset of mitosis could not be retarded by fusion with a younger plasmodium (Sachsenmaier, 1981).

For eukaryotic cells, therefore, with nonlimiting provision of nutrients, the attainment of a minimum critical size is an essential but not sufficient factor before DNA synthesis is initiated. The nature of the critical size-dependent requirement is uncertain, controversial, and possibly different for different cell types. Physical factors dependent on increasingly divergent volume:surface ratios can be interpreted to favor dilution of inhibitors or increases in amount of essential intermediate/cell. Cell surface:nuclear size ratios may be relevant in the light of current developments in microstructure (see Chapter 6) but have not yet been much considered. Sufficient energy (high ATP:ADP), rapid rates of protein synthesis, provision of essential intermediates for replication, and essential structures for nuclear division have all been proposed as limiting factors (see Chapter 7). Specific gene products are certainly required at defined periods before the onset of S phase and mitosis. Interpretational difficulties can arise from delays between transcription, translation, and intracellular actions. The size element in G_2 is uncertain, and normally additional specific protein synthesis is needed before mitosis. In mammalian cells, unlike those in yeasts, DNA replication, mitosis, and cell division cycles are simply linked (see Lloyd et al., 1982).

Experimentally determined rates of growth through the cycle vary considerably in different cell types (see Prescott, 1976). In *Amoeba proteus*, which has no G_1, the rate of growth declines through G_2. In fibroblasts, conversely, rates of protein synthesis increase as the cycle progresses. In early cleavage stages in embryos when there is no G_1, synchronous division occurs through several cell cycles without growth or new RNA synthesis. S phase is also considerably shorter, probably because more replicatory origins are in use than in adult cells.

QUIESCENT CELLS—THE G_0 PHASE

Most adult somatic cells with which regeneration is normally concerned are not continuously traversing the cell cycle just discussed. They are in an extended, sometimes very prolonged, G_1 phase or, as first suggested by Lajtha (1963), in G_0 phase (see also Baserga, 1976). Operationally cells in G_0 may be recognized by some or all of the following functional characteristics:

1. Extracellular factors additional to normal nutrients are required to promote the cells into replication and division.

2. The transit time for the first G_1 phase through which the population passes after stimulation is significantly longer than the G_1 phase for a second cycle (see Baserga, 1976).

3. Normally the permeability properties of the plasma membrane of cells in G_0 are constrained.

4. Proportions of nuclear nonhistone proteins to DNA are frequently lower.

5. Transcriptional activity in chromatin is reduced.

The greater the time spent in G_0, the longer the time taken after stimulation to reach S phase (see Baserga, 1976).

Reconciliation of these observations with those from unicellular organisms, yeasts, and cleavage-stage embryos suggest that in an archetypal division cycle G_1 phase would not exist (D. M. Prescott, unpublished) (see Fig. 1.1).

Provided they are large enough, cells would move out of mitosis and immediately initiate DNA synthesis, followed by G_2. Such a concept forces consideration of classes of quiescent cells, of which the extreme, physiologically irreversible class would be exemplified by the neuron. Regeneration in higher animals commonly involves the promotion of cells from quiescence in G_0 into active cycles.

GROWTH FACTORS (GOSPODAROWICZ & MORAN, 1976)

One final point which arises from consideration of growth in eukaryotic cell cultures is the importance of growth factors in maintaining the cultures in a proliferating state. Excluding the provision of adequate energy sources and essential nutrients, two types of activity are evident—that which is required to maintain the viability and phenotypic characteristics of the culture and that which is specifically essential for growth and division of the target cells. Growth factors in these senses will be considered in the sections on signals (Chapter 4) and on the membrane (Chapter 6).

1.1.3. Mathematical Models

Many attempts have been made to interpret growth and morphogenesis in mathematical or physicomathematical terms. In this century D'Arcy Thompson's seminal essay "On Growth and Form" (1917) stimulated two main lines of analysis relevant to our purposes—those predicting cell cycle time and the parameters which regulate this, and those attempting to describe the origins of pattern formation or morphogenetic fields.

Cell Cycle Analysis

Two main theories have been advanced to interpret the observed relation between cell size and population growth—a deterministic model in which the attainment of some critical parameter is a prerequisite for DNA synthesis and cell division (e.g., Wheals & Silverman, 1982), and a probabilistic model which was introduced to explain the observed differences in cell cycle times (coefficients of variation 10–20%) between sister cells derived from the same clone in the same extracellular environment (Smith & Martin, 1973). Developments of transition probability models (Shields, 1977; Shields et al., 1978), which include both deterministic elements related to cell size as well as transition probabilities, allow predictions of cycle times which are very close to experimental observations. Sophisticated computer techniques enable experimental data to be tested against a number of alternative theoretical curves, but the differences required to discriminate between the different models may be less than can be attained by experimental observations (Smith et al., 1981). At the simplest level current analyses are consistent with an A state in G_1 from which cells escape into a second state, B, with constant probability/unit time. The B state comprises part of G_1, S, G_2, and M. Transit times through S, G_2, and M are thought to be invariant, and the part of G_1 in B is the critical variable. A model of this sort is applicable under conditions of rapid growth. Complications have, however, arisen in adapting the models to situations when rapid growth can no longer be maintained. Difficulties are found especially in accommodating experimentally observed lags which precede increases in proliferation rate after mitotic stimulation, for example, following the addition of fresh serum to 3T3 cells (Brooks, 1981; Hirsch, 1983). Once out of the quiescent state, rates of transit do not appear to differ between cells which were previously quiescent and those which were in normal growth cycles (Brooks et al., 1980).

With our present uncertainties in identifying limiting parameters in simple growth cycles, it is impossible yet even to conclude that deterministic strategies are identical for pro- and eukaryotic cells. Nurse (1980) suggests that in bacterial cell growth the probabilistic transit is very rapid and the deterministic element dominates, with the relative importance of the two states reversed in mammalian cells.

Promotion out of a quiescent state into active growth is more relevant to regeneration than considerations derived from logarithmically growing cell cultures. Regeneration in most multicellular organisms occurs against a well-orchestrated hormonal background, including both stimulatory and inhibitory factors. Further complications arise if circadian rhythms influence the organism within which growth has been selectively stimulated. The production of such rhythms by feedback controlled processes can be accommodated into transition probability models so that, when interactions occur between periodic events and a (linearly) developing system, accurately fitting analyses emerge (see Goodwin, 1976).

1.2. UNICELLULAR ORGANISMS

Some unicellular organisms, such as *Bursaria truncata,* a ciliated protozoan, can be divided into two and, provided both parts contain a portion of the very high-ploidy macronucleus, they are capable of reconstituting a viable organism (Lund, 1917). Unicellular organisms with diploid nuclei cannot regenerate if their nuclei are divided. The fully grown *Bursaria* has a well-developed morphology with a ciliated adoral zone, mouth, and gullet; these disappear and are reformed in the regenerative process.

A feature of the behavior of *Bursaria* which will recur in many of the multicellular systems to be considered is the disappearance of the characteristic morphology of the adult organism before restoration of the ablated regions commences. *Dedifferentiation* is a commonly used description of this event, conveying several meanings. One is a functional or morphological change in the physiological activity of the cell so that properties characteristic of an earlier stage in development are reacquired. This sense was defined as *modulation* by Weiss (see Schmidt, 1968; Weiss, 1973); it frequently involves a resumption of mitotic activity, migration of previously stationary cells, or similar manifestations of features of cell activity more evident in the embryonic than the adult state. Gene functions which are characteristic of the differentiated somatic state of the cell may or may not continue to be expressed, and others are elicited which are periodically exhibited in the life cycle of the particular type (facultative expression).

A second sense of dedifferentiation is metaplasia—a change in cell form (and function). Cells dedifferentiate and subsequently bring into play a new family of genes characteristic of a different cell type within the same embryological lineage; for example, fibroblasts can become chondrocytes or osteocytes.

The concept of dedifferentiation is extremely controversial (Weiss, 1973; Truman, 1974); many authorities refute the idea that differentiated cells can change their behavior and later exhibit and transmit to daughter cells different characteristics. The nuclear transplantation experiments of Gurdon and others (Section 1.1.2) may be interpreted to demonstrate the capacity of the nucleus for redifferentiation when introduced into a different cytoplasmic environment. The intact cell, with its interdependent nuclear-cytoplasmic system interacting homeostatically as a buffer against external perturbation, is often thought incapable of such a response (but see Truman, 1974). Proponents of this argument postulate the presence of stem cells for repopulation (see Section 3.2). Possible stages in modulation experienced by a cell as it becomes differentiated, and the probability of different strategies for differentiation, some of which produce irreversible changes in chromatin (Section 8.2.3), blur distinctions in this area.

1.3.   REGENERATION IN MULTICELLULAR ORGANISMS: INVERTEBRATES

Complete regeneration of a multicellular organism from its parts occurs only in invertebrates, notably the hydroids and worms (see Hay, 1966; Rose, 1970). Regeneration of complete limbs or tail occurs in limited orders of vertebrates up to Amphibia. The urodeles (newts, salamanders) are the best-studied examples and contrast with anurans (frogs), since the latter do not normally, as adults, reform amputated limbs. We shall consider here only limited examples of regeneration to illustrate basic principles for which a biochemical explanation is still to be provided.

1.3.1.   The Hydroids

The invertebrates most commonly used are *Hydra* (fresh water) and *Tubularia* (marine). Regeneration in *Hydra* was first studied by Trembley (1744). The tips of the tentacles and the basal disc cannot be used for regeneration; otherwise any part of the organism can be reconstructed from small fragments of the gastric column (see Fig. 1.2). This capacity is also shown by a section, and only if the section is very thin (< 2 mm) does abnormal de-

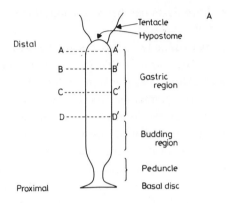

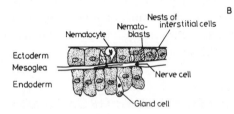

Figure 1.2. Hydra. (A) Schematic diagram (see also Table 1.1). (B) Detail of cell layers (Gierer). Reprinted by permission of Academic Press (*Current Topics in Developmental Biology*, Vol. 11, p. 19, 1977).

velopment occur to give head regions from both the cut surfaces. If cells from a midsection (Fig. 1.2) are dissociated in an appropriate medium, they will reaggregate, reestablishing the normal radial distribution of ectoderm and endoderm, and subsequently regenerate a complete organism (Gierer, 1977). Further, if cells from the head area are dissociated, the regenerate will have large numbers of tentacles, and similarly a preparation from the foot section will mainly form feet.

The type of regeneration is morphallactic and bidirectional, that is, new parts are restructured from preexisting differentiated cells and potentially can regenerate both distal and proximal structures. Bidirectional regeneration is shown only by organisms which undergo asexual reproduction by a method which itself involves regeneration as an integral part of the process (e.g., in *Hydra* by transverse fission following budding) (Slack, 1980a).

Three features of regeneration in the hydroids have especially been studied—the multipotency of the cells involved, the polarity of the process, and the agents which can mimic or influence polarization. Ectodermal intestinal cells are multipotent, on average producing about 50% of stem cells, 10% of nerve cells, and 30% of nematocyte precursors (see Gierer,

1977; Heimfeld & Bode, 1981). Which ultimate fate ensues is determined by the location in the animal; if the interstitial cells are situated in the head region, nerve cells predominate, whereas in the gastric region nematocytes are more evident. Nematocytes can migrate into the tentacles, in response to signals from the head region, along a complex route which may also be "signposted" (Gierer, 1977) (Sections 6.4.2 & 6.5).

As will be apparent whenever regeneration or hyperplasia occurs, regulatory mechanisms must be operating to balance the need to maintain adequate numbers of proliferating stem cells (in *Hydra,* interstitial cells) and the requirement for nondividing highly differentiated cells. The system will include some mechanism by which cell depletion is signaled back to the stem cell, causing it to divide to yield at least one daughter cell which moves out of the division cycle and differentiates further.

Two lines of evidence suggest interstitial cells are not the sole contributors to the regenerate. Classical experiments (see Rose, 1970), with *Hydra* having its endoderm marked by the presence of chlorophyll-containing symbionts, showed that this epithelium could regenerate the whole organism. Additionally, if the animals were exposed transiently to methylene blue, the redox dye remained oxidized in the endoderm, staining it blue, but the interstitial cells in the ectoderm were colorless. If the *Hydra* was then amputated at, say, BB' (see Fig. 1.2), the regenerated head region contained endoderm which was still blue, showing that this tissue had contributed to the regenerate.

Polarity was evident in all the classical amputation studies; the residual portion regenerated to form more distal parts not already distal to it (Rose, 1970). Further, the influence of the head region was dominant, since the presence of an existing head inhibited the generation of a head from a piece of head grafted immediately adjacent to the preexisting head region. This inhibitory effect of head disappears by the budding region. In *Tubularia* if the coelom was blocked by an oil drop, head dominance was also blocked. There is a similar situation with foot transplants, and it is suggested that different inhibitory systems exist for the two cases.

Polarity in the hydroids is associated with an externally imposed asymmetry (Zwilling, 1939). This can be mimicked by a local increase in the partial pressure of oxygen. Selective exposure of the proximal surface BB' of a section AA' BB' (see Fig. 1.2) can reverse its original polarity and promote head regeneration from the surface BB'. Selectively increasing the temperature of one surface produces a similar effect. Conversely, treatment

TABLE 1.1.   REGENERATION RESPONSES IN HYDROIDS

Surgical Procedure	Response
1. Single section	Regeneration of the part removed, except if tips of tentacle or basal disc taken.
2. Disc (sectioned at AA' and BB')	Regeneration of whole organism, hypostome regenerated from surface AA'
3. Sectioned at AA', BB', and CC', grafted to give BB' AA' CC'	Hypostome formed from BB'
4. Sectioned at CC' Sectioned at AA' and CC'	Foot regeneration in about 4 h Foot regeneration in 1 h
5. 3 sections created: AA' BB', BB' CC', CC' DD', and ends ligatured	Rate regeneration of head and foot faster in AA' BB' than BB' CC' than CC' DD'

Note: Sections shown in Fig. 1.2.

with reducing agents leads to the formation of "foot." The formation of heads at both ends of very thin sections of the gastric region is therefore due to the absence of asymmetry in the external environment, and thus to a failure of one or other surface to develop preferentially and produces an inhibitor preventing head formation from the contralateral surface.

Barth (1938, and see Wolpert, 1971) showed rates of regeneration to be slower the more proximal the regenerating section (see Table 1.1). In time, however, each section formed a complete regenerate, and it would appear that head formation is associated with, but not dependent upon, rapid growth.

Regulatory factors promoting or inhibiting head and foot formation have been isolated. The head stimulator is a polypeptide (M_r c. 1 kDa) (Schaller, 1973) which increases the tentacle number in the regenerate and was thought to be effective at about 0.1 nM. Its concentration was correlated with the presence of nerve cells and was thus enhanced intracellularly in the head region. An inhibitor blocking the development of head (M_r c. 0.5 kDa) (Berking, 1977) was also found in the head region. It was formed only when there was high metabolic and mitotic activity and might be a catabolite. The inhibitor was activated when released extracellularly and could diffuse out from the organisms. Its concentration diminished down the stalk so that the tentacle-stimulating agent became the prominent determining factor by the budding region. The foot inhibitor (M_r c. 0.5 kDa) (Schmidt & Schaller, 1980) was also present in the nerve cells and was neither a peptide (resistant to 6 M HCl at 100°C for 48 h) nor a known neurotransmitter. Similar experiments have been performed with *Tubularia* (see Rose, 1970), in which electrophoretic separation of proteins from homogenates of the distal region yielded positively charged peptides inhibiting distal development. They were believed to act as repressors blocking development downstream from their sites of origin.

1.3.2. Planaria and Other Worms (see Rose, 1970; Brønsted, 1969)

Regeneration by *Planarium*, a hermaphrodite nonsegmented flatworm (see Fig. 1.3), was intensively studied by Morgan and Child at the beginning of this century and by Brønsted more recently. The regeneration is bidirectional and epimorphic (cell division precedes reformation of structure). In contrast to *Hydra*, following amputation, a region—the blastema—is formed under a covering of epidermis. The cells of the blastema lose many of their morphological and biochemical characteristics before starting to divide and ultimately redifferentiating (Sections 1.2 & 8.2.3) to form the regenerate. Regeneration by *Planarium* shows polarity, head dominance, and a graded response down the body and is big enough for some biochemical studies.

Regeneration can take place from all regions of the organism except parts in front of the eyes, but reformation of the eyes is obligatorily dependent on the prior presence of head ganglia, although little or no information is available on the processes involved. The rate of reformation of head, monitored by eyespot formation, slows down toward the tail and is very slow for sections behind the pharynx. If the animal is sectioned at AA' and 12 h later cut at BB' (see Fig. 1.3), regeneration at BB' is promoted. Also, if it is cut at AA' and regrafted to the original tail section, no cells are infilled,

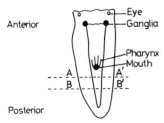

Anterior

Posterior

Figure 1.3.   Planarium (see Section 1.3.2).

but if it is cut at AA' and BB' and the two amputation planes are juxtaposed, cells are regenerated to fill in the normal section.

The cells of the blastema are derived from cells immediately adjacent to the amputation surface and/or from ones which migrate into the wound from greater depths. Labelling with ^{3}H thymidine (TdR) indicates that they commence proliferation just under the surface of the wound. Details of this process are controversial; it is not known whether all somatic cells can dedifferentiate (Sections 1.2 & 8.2.3) and contribute to the blastema, or what causes their migration to the site and promotes mitosis (see Chapters 2 & 6). The *neoblast* theory alternatively proposes the blastema is formed exclusively from embryonic reserve cells (cf. stem cells, Section 3.2.1) that are undifferentiated and widely distributed in the adult.

With *Dugesia lugubris*, mosaic individuals have been constructed from maternal hexaploid animals (6n = 24) and paternal diploids (2n = 8) to give progeny with triploid somatic cells (3n = 12) (Gremigni & Miceli, 1980). If regeneration is promoted in a gonadless region, the regenerate contains somatic tissues which are wholly triploid. If the amputation plane is in the region of the gonads, the blastemas contain diploid and hexaploid cells (see Slack, 1980b), indicating that highly differentiated cells in these animals remain multipotent and can give daughter cells of quite different phenotypes. Respecification of the phenotypes of the cells in the blastema is required and it seems unlikely that all the spatial information from their previous site is conserved.

New instructions to the cells of the blastema are conveyed very early. Some of this information is already present in the blastema as soon as it can be separated from the body (three days), since an early blastema can then form a head or a tail but his tiny part cannot develop further to make a complete animal. It appears that, at this time, only the minimum of instruction is available to make eyes, brain, and muscle (Sengel, 1960).

As with *Hydra* the polarity of the regenerate can be altered experimentally. If sections are cut very thinly, the anterior-posterior polarity may not be established. Deacetylmethylcolchicine (colcemid) (Kanatani, 1958; Flickinger & Coward, 1962) caused surfaces, which would otherwise have formed tails, to form heads, so bicephalic regenerates were obtained. If glucose was added, the effects of the colchicine were prevented.

The existence of a natural metabolic gradient was suggested by Child and demonstrated in the platyhelminth *Dugesia tigrina* by Flickinger (1959). Uptake of $^{14}CO_2$ or 1-^{14}C glycine into total protein diminished through the body from head to tail. If this metabolic differential between head and tail was upset by exposure of the anterior portion of the worm to various inhibitors, so reducing the production of protein relative to the rest of the body, amputation followed by reexposure of the head region to the drug sometimes produced a second head where the tail should have developed (Flickinger & Coward, 1962).

Other experiments have examined the possibility that polarity might be associated with electrical gradients through the animal. Midbody sections of *D. tigrina* were placed onto agar, and potential differences were applied across the sections for five days. With weak fields the polarity of the applied potential had no effect, but if the field exceeded about 20 μA/mm^2, and the potential 225 mV/mm, the polarity of the regenerate could be permanently reversed (see Rose, 1970). The role of naturally produced electric fields in regeneration has been reviewed recently (Borgens, 1982).

A further feature of the regeneration shown by segmented worms is that, when posterior segments are amputated, replacement occurs to reestablish the original number of segments (see Rose, 1970). In *Eisenia* the posterior tip of the worm is electropositive with respect to the rest of the body. Following amputation, segments are replaced until the summation of the electrical potential in the added segments restores the critical inhibitory voltage for further growth (Moment, 1953).

1.3.3. Summarizing Conclusions

In spite of the elegance of classical work on regeneration in worms, it is evident that we know virtually nothing about the biochemical details. Morgan (1901, 1904) and Child (1941) both directed attention to the essentiality of preserving part of the nerve chain in earthworms if regeneration was to occur. What happens to allow reformation of the head ganglia in *Planarium* is unknown. Further, when nematode development has been studied, cell interactions have been essential, although the relation between the induced or inducing parts is not yet clear (Sulston & White, 1980). While frequent and often useful comparisons are made between development and regeneration, similarities are not complete (see Chapter 2), and processes in development in one order do not necessarily allow predictions about strategies in another.

A further major gap in the work from both hydroids and worms is that regeneration requires three-dimensional (3-D) information, but head-to-tail polarity is the only axis about which data are available. These appear to indicate an oxygen-nutrient-metabolic gradient of unknown form which diminishes from head to tail. Its existence could be either causative or consequential to head determination.

The needs for growth and for trophic factors for its promotion will emerge as common features in regeneration and compensatory hyperplasia. ATP availability, redox differences, and high rates of protein synthesis can have widespread biochemical consequences but are informationally nonspecific. The most that might be postulated is that, for head formation to occur, the anabolic potential of the cells involved is maximally extended. The signals allowing respecification of the cells and the possibility that these are simultaneously directional are quite unknown. The various biological theories that have been advanced to produce 3-D information will be considered after limb regeneration has been reviewed.

2

THE SYSTEMS— VERTEBRATES

2.1.   REGENERATION OF URODELE LIMBS (Schmidt, 1968; Wallace, 1981)

Aristotle (c. 330 BC) was probably the first to record that regeneration took place after a lizard lost part of its tail. In 1769 Spallanzi reported that newt limbs could regenerate. This capacity is shown for many structures in tailed Amphibia but not in adult anurans. Excellent comprehensive accounts of regeneration through the animal kingdom have been given by Hay (1966), Goss (1969), and Rose (1970). We shall discuss here only a limited number of examples of regenerating systems, chosen to illustrate biological principles and the problems which may be involved. Mammalian tissues such as the liver and kidney, which show compensatory hyperplasia after injury, will be considered in Chapter 3, as these have been intensively studied at the molecular level (see Chapters 6, 7, & 8). Skin repair and the restoration or stimulation of bone marrow function will also be included because of the information they are providing about the control of growth and proliferation in multicellular organisms and about genetic factors which may limit regeneration in some cell lines.

2.1.1.   Normal Regeneration of Urodele Forelimbs

After amputation blood from the severed vessels clots, and the vessels constrict to reduce the blood flow. Blood lost in urodele amputations is minimal. Epidermal cells at the periphery of the wound become less adherent to one another and migrate over the surface of the wound to close it off. The cells accumulate near the middle of the surface to form a thickened apical cap. Beneath this cap the cells of mesodermal origin dedifferentiate (Sections 1.2 & 8.2.3) and accumulate to form a blastema. The blastema is composed of a morphologically homogeneous mass of cells which increase in number at first by the addition of more cells and later, at about 10 days, as shown by ^{3}H thymidine incorporation, by cell division. Restructuring the wound area commences through migration into the region of macrophages and other white cells, perhaps promoted by chemotactic agents released from the damaged tissues (Section 6.5). The region becomes relatively anoxic, the pH falls to 6.9–6.7, and edema is evident. The fall in pH facilitates the activity of some lysosomal hydrolases so that the damaged tissue is eliminated. Certain of the enzymes of the tricarboxylic acid cycle are lost or are present in only very low amounts (Schmidt, 1968). When the differentiation of the blastema cells commences, the lactic dehydrogenase activity subsides and the Krebs cycle takes over again.

The blastema is completely formed by about 15 days; its appearance is conical at 20 days and it is overlain by the apical epidermal cap. At this time the cells in the proximal region of the blastema begin to differentiate while the undifferentiated cells in the distal region continue to divide. The blastema then flattens to become paddle-shaped, two to three digits become detectable at about 30 days, and four digits by 35–40 days.

THE ORIGINS OF THE BLASTEMAL CELLS

If [3]HTdR is administered just before amputation, cells which are in S phase at the time of the operation will have labelled DNA. In normal adult amphibian limbs only epidermal cells are so marked. After amputation, labelling studies confirm the migration of the epidermal cells over the wound surface. Apical epidermal cells retain a constant amount of label/cell, but cells immediately proximal to the amputation plane show dilution of the label as the number of mitoses rises. These cells migrate distally, replacing those in the apical cap; they are then sloughed off, as for epidermis in normal skin.

As the blastema forms, its cells are largely unlabelled, confirming suggestions from the older histological experiments that mesenchymal cells in the blastema originate from internal tissues with no contribution from the cells of the epidermis.

Labelling with [3]HTdR has a number of drawbacks. Only that fraction of the population (not greater than 25%) of the cells in S at the time of the exposure incorporates the label. Since free thymidine is usually rapidly destroyed, the period for which the marker is available is normally substantially shorter than the duration of the S period, so only a small proportion of the proliferating population can get labelled. If division ceases, the label is relatively stable—repair replication (unscheduled DNA synthesis) may be minimal; but if proliferation is active, successive divisions rapidly reduce the level of the marker to background. Alternative methods of cell recognition are available and especially suitable if blastemal grafts are in use. These can be chromosomal markers or differences in ploidy. If epidermal recognition is sufficient, skin color may be used, and sometimes differences in morphology are adequate. Current techniques of handling animals carrying the regenerate now enable statistically significant numbers of animals to be examined and their cells to be identified microscopically.

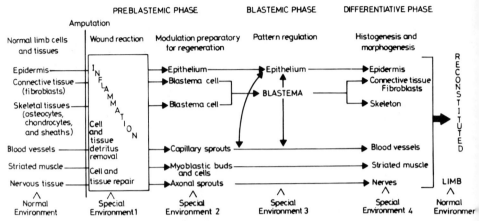

Figure 2.1.   A concise summary of the regeneration process in amphibian appendages (Schmidt). Reprinted by permission of University of Chicago Press (*Cellular Biology of Vertebrate Regeneration and Repair*, p. 306, 1968).

2.1.2.   Tissue Replacement and Cell Interactions (Fig. 2.1)

Once the blastema has been established, cell proliferation, the selection of the required set of phenotypic characteristics, and growth have all to occur in the correct temporal sequence to give rise to appropriate tissues in the right places in the regenerate to restore size, shape, and function of the limb with its correct symmetry. Similar processes had also to occur during embryonic development; most biologists have presumed that processes of development and regeneration are very similar, although differences in detail are certainly apparent. We shall consider first how the tissue responds and then the theories of how these responses were achieved.

The Epidermal Changes

These have already been described. Classical experiments (see Schmidt, 1968) established that the presence of an apical epidermal cap was essential— its careful, repeated removal completely stopped regeneration (Thornton, 1957). Innervation of the epidermis occurs very early (c. 48 h) in the regeneration process. If the stump is surgically manipulated to interpose dermis under the epidermal layer to prevent epidermal contacts with axonal outgrowths, regeneration is blocked.

X-irradiation stops cell proliferation, and irradiation of the stump and/ or blastema has been intensively used in studies of regeneration. Exposing a few mm of the stump to 20 Gy prevents regeneration because the cells are no longer able to divide and form the blastema. From this it was concluded the cells of the blastema were of local origin (see Butler, 1933; Wallace, 1981).

Interactions between the stump and the blastema were investigated by Holder and colleagues (1979), who grafted a right or left blastema onto a left or right stump, respectively. The control experiment with the handedness reversed gave good regeneration, but if the blastema was irradiated with 20 Gy, the regeneration was very poor and often regressed: in 10 of 27 cases the handedness was normal, showing the blastemal graft had died. Similar experiments grafting a tail blastema onto a limb stump yielded a tail-limb combination, but if the stump was irradiated only the tail developed (see Table 2.1). It is likely that the irradiation had caused the release of hydrolases and of phosphatases from the lysosomes, with consequent cell lysis and failure of regeneration.

Uptake of ^{3}HTdR into epidermis was reported to continue in irradiated *Triturus* limb (Rose & Rose, 1965), suggesting that the epidermal layer was unaffected. Regeneration, however, did not occur. Unscheduled DNA synthesis is more radioresistant than cell division, so damage to the epidermis cannot be ruled out. If cuffs of nonirradiated skin were applied to irradiated stumps, with the normal orientation preserved, regeneration was again found (Bryant, 1976), confirming the importance of epidermis in the process (see Chapters 6 & 9).

NERVE REGROWTH

The importance of nerve is one of the distinguishing features between development and regeneration. As the limb bud develops on the flank of the embryo, it is covered with putative epidermis, but invasion from the nerve occurs late, after the future of the cells has been determined. If, however, adult limb is denervated at the time of operation, limb regeneration is prevented (see Wallace, 1981), but if denervation takes place after the blastema is formed, regeneration still takes place but to a lesser extent. Singer (1952), using newts, tried to find promoter substances without success; acetylcholine was not the factor.

If the brachial plexus is transected, only the cut axon degenerates, not the Schwann cells, but regeneration of the limb does not occur until the

TABLE 2.1.   SOME PATTERNS OF REGENERATION IN URODELE LIMBS

	Stump		Blastema	Result
1.	Normal		Normal	Normal regenerate
2.	Normal	a.	Graft from "established blastema" AP axis reversed *or* from contralateral side	Disharmonic development— partial duplication
		b.	Ipsilateral side, normal blastema grafted back	Normal regeneration
		c.	Very young contralateral blastemal graft	Normal regeneration
3. a.	Proximal to blastema		Ipsilateral blastema graft	Intercalary regeneration
b.	Distal to blastema		Ipsilateral blastemal graft	Regeneration with duplicated structures
4.	Irradiated	a.	Ipsilateral blastemal graft	Regeneration
		b.	Irradiated blastemal graft	No regeneration
5. a.	Irradiated limb stump		Tail blastemal graft	Tail regeneration
b.	Irradiated tail stump		Limb blastemal graft	Limb regenerated

axon regenerates. The capacity to support regeneration is shown by both motor and sensory nerves and appears to be directly related to the proportion of axons remaining in the partially denervated limb. If the limb was denervated soon after the animal hatched, the limb was nonfunctional but would reform after amputation. This shows the nerve axons are not always essential for tissue reformation and organization in adult urodele limbs.

Following amputation of a normal limb, retrograde degeneration of the myelinated nerve axons is detected early, possibly restricted to motor nerves (Schmidt, 1968). Schwann cells from the nerve sheaths migrate towards the apical wound epithelium and become engorged with lipid from broken-down cells and myelin. The migrating cells serve as guidelines for the regenerating axons. Tissue culture experiments have shown the essentiality of this guidance to the regenerating axon if proper nerve function is to be restored (Johnston & Wessells, 1980; Solomon, 1981). Adhesion to the substratum on which the axon is growing appears to be essential in culture, implicating glycosoaminoglycans on the Schwann cells for their guidance (Section 6.4.2). Interdependence between the tip of the axon and the muscle which it will innervate is well established embryologically (see Varon & Bunge, 1978). It is quite unclear the extent to which the properties of the cells in the blastema and their immediate extracellular environment can properly be equated to the situation in the embryo. Especially is this true with respect to the glycosylated residues on their cell surfaces and the regulatory molecules which may interact with them.

VASCULAR REGENERATION

Little is known of the details of the restoration of the circulatory system to the regenerating limb. It is assumed that the operating processes are similar to those shown in the restoration of circulation in mammalian injuries. Here endothelial cells in capillaries adjacent to the wound divide, and sprouts arise which fuse with others to form loops which then become patent, allowing blood to circulate. Endothelial cells involved in the sprouting have more evident endoplasmic reticulum than normal endothelial cells; mitoses are seen in cells behind the growing sprout tips. Initially the system is loosely organized; the direction of blood flow in a recently formed loop can be reversed, and if blood does not flow into the loop, it is broken down. As tissue develops surrounding the vessels, irregularities in the pressure arise and, in the regions of higher pressure, endothelial cells somehow acquire muscle coverage to form arterioles.

MUSCLE, CARTILAGE, CONNECTIVE TISSUE, AND BONE (SEE SCHMIDT, 1968)

These tissues of mesenchymal origin form part of a large group of cells whose pathways and development diverge from one another at slightly different stages of embryonic life. In many cases the different developmental end points can be altered experimentally by the presence of apparently unrelated *inducing* cells, for example, the induction of ectopic bone by transitional epithelium (Friedenstein, 1968) (Section 6.4.2). In other cases defined chemical changes bring about the effects. In chick wing bud cultures, for example, Caplan (1981) has shown that, under conditions of high cell density, cartilage-forming chondrocytes appear; with medium density, bone is formed, and at low density, muscle cells. The wing bud is relatively avascular in early development, the central region—destined to become cartilage or bone—having a limited flow of nutrients. High nicotinamide adenine dinucleotide (NAD) levels, derived from a relatively ample supply of tryptophan, are thought by Caplan to be involved in switching on processes leading to muscle protein synthesis in the better vascularized, peripheral, premyogenic areas. Low NAD levels are associated with the presumptive chondrocytes. The lability of protein expression by these mesenchymal cells is the reason behind the suggestion that, in the blastema, the required specifications for the new patterns of development could be determined by agents similar to those employed in the embryo (Mohun et al., 1980).

Respecification is probably not required for muscle in normal regeneration. Myofibrils in adult striated muscle are present in terminally differentiated multinuclear myotubules, which are thought not to be capable of division. It is believed that muscle in the normal regenerate arises principally from mononuclear myoblasts or presumptive myoblasts in the muscles of the stump adjacent to the amputation plane, which are already committed to muscle protein synthesis. The myoblasts become aligned, fuse, and synthesize muscle proteins (Bintliff & Walker, 1960; Dienstman & Holtzer, 1975; Merlie et al., 1977; Holder, 1981). When regeneration occurs with an irradiated stump over which a normal blastema has been grafted, the origin of the muscle tissue is not settled and may involve respecification of cells in the blastema (Desselle & Gontcharoff, 1978).

As with nerve axons, elongation of the myotubules may necessitate their ordered alignment, which in tissue culture studies can be provided by adhesion to fibronectin (Chiquet et al., 1981) (Section 6.4.1). The interdependence of nerve, muscle, and bone growth on normal limb function has already been noted.

As a result of amputation, chondrocytes and osteocytes are released from the exposed lacunae in the bone. A cell type from earlier in their lineage—the fibroblast—becomes manifest. Fibroblasts are thought to be as ubiquitous in Amphibia as in mammals and are a major contributor to the cells in the blastema (Schmidt, 1968). In mammalian injury, collagen formation and deposition in granulation tissue (the equivalent of the blastema) is very prominent, and repair, rather than regeneration, is the normal outcome (Bourne, 1981; Roels, 1981). In urodeles cell erosion into the limb stump ceases when chondrocytes within the periosteum proliferate and form a callus around the site of the injury. In normal mammalian and avian growth, parathormone and thyrocalcitonin are involved in bone formation, affecting the balance between deposition by osteoblasts and resorption by osteoclasts (Owen, 1970; Moss, 1972). New deposition of bone occurs on the surface of the callus. The microenvironment, especially the partial pressure of O_2, is critical in the balance between osteogenesis and resorption (Fell, 1969). Induction of bone deposition occurs rather easily, including by bone itself. Stress consequent on renewed weight bearing is a further important factor surrounding bone shape. These observations again focus on the importance of local factors in specifying the phenotypic behavior to be shown by fibroblast related cells.

2.2.   MODELS FOR LIMB REGENERATION

Almost all analyses start with embryonic limb development, and most with the developing chick wing bud since this embryo has long been open to experimentation (see Ham & Veomett, 1980). Before, however, considering the limb system, some results from developmental studies in *Drosophila* embryos must be briefly reviewed because of the power of the genetic analyses available here and the influence this approach is exerting on biochemical interpretations of morphogenesis.

2.2.1.   Drosophila Development (Gehring & Nöthiger, 1973; Ham & Veomett, 1980)

Up to 13 cleavage stages occur in *Drosophila* without cell division, so a syncytium with synchronously dividing nuclei is formed. By the 13th cleavage the nuclei have moved outwards to the periphery of the embryo, nuclear division ceases, and membrane separation occurs to give the cellular blastoderm. At this stage the cells of the blastoderm can be separated into two groups: those associated with the anterior end of the embryo, marked by a micropyle, and those from the posterior end around the polar cells.

If the embryos are obtained from morphologically recognizable individuals—yellow or ebony, forked bristles, or multiple wing hairs—the cells can subsequently be recognized in wild-type hosts. Disaggregated cells from the anterior or posterior halves of the blastoderm are mixed with oppositely marked cells from whole blastoderms and introduced into wild-type adult hosts, in which they proliferate. When sufficient cells are available, they are put into metamorphosing (third instar) larvae to differentiate. The consequent structures can then be identified morphologically and ascribed uniquely to the anterior or posterior region of the embryo.

At later developmental stages sequential formation of lines of cells which are clonally restricted can be studied in embryos which are heterozygous for some morphologically recognizable recessive trait, for example, multiple wing hairs (mwh) (Kauffman et al., 1978). If larvae are irradiated, somatic recombination can occur in G_2-stage cells to give two daughter cells, one of which is homozygous for the recessive character (see Ham & Veomett, 1980).

Four aspects of *Drosophila* embryology are especially relevant:

1.   The very early determination (3–10 h development) of anterior-posterior zones in the embryo (Chan & Gehring, 1971), followed thereafter by the dorsoventral and proximo-distal axes (see Kauffman et al., 1978). Commitment of the cells is autonomous and transmissible to the next generation. Anterior or posterior properties are not affected by the admixture of blastodermal cells with different presumptive fates.

2.   A second influential finding was that of homoeotic mutants—those which alter the properties of an imaginal disc or disc region so that it is committed to a fate normally shown by another disc (see Garcia-Bellido & Ripoll, 1978). Such mutants include *bithorax* and *Nasobemia* (Ham & Veomett, 1980), in which the adult carries a leg in the antennal position. Mutants of this sort, where one structure is replaced by another virtually intact part, suggest genetic control over complete patterns of development. In some cases the effects of the mutant are restricted. Engrailed (en)—a mutant which causes posterior structures to behave as anterior (see Ham & Veomett, 1980) and which affects early development in *Drosophila*—is only effective in cells which are in posterior compartments. If en clones are present in the anterior compartment, development is normal. Other homoeotic mutants show similar restrictions (Morata & Kerridge, 1982).

3.   Imaginal discs transplanted through adult hosts and into larvae for metamorphosis may show transdetermination, that is, unexpected differentiated structures appear. These are not random but show clear lineages (genital → antenna ⇆ wing → thorax) (Ham & Veomett, 1980). Analyses of these transformations and of homoeotic mutants have led Kauffman and colleagues (1978) to suggest that development proceeds through a series of binary choices (on/off), with the presence of a morphogen at a concentration

exceeding a threshold allowing a program of protein synthesis, leading to a particular structure, to be switched on. Critical morphogen concentrations were thought to arise by a reaction-diffusion mechanism (see following).

4.   A further point emerging from the *Drosophila* studies is the evidence that commitment, both functional and positional, is accompanied by the appearance of cell surface markers, which allow intercellular recognition, and the adhesion of homologous cells (Garcia-Bellido, 1966). Marked cells from imaginal discs were dissociated, mixed, and allowed to develop through adult and metamorphosing hosts. If cells from different discs were mixed, heterotypic cells sorted out so only cells derived from the same disc associated together. If marked cells from wing discs were mixed, homotypic but chimeric associations were seen. More recent experiments with immunochemical techniques demonstrated that an antigen initially present on all epithelial cells became restricted to cells derived from the dorsal region of the disc. The presence of the antigen was unrestricted with respect to the structure that finally developed, that is, the antigen was position specific (Wilcox et al., 1981).

2.2.2.   Limb Bud Development

Limb bud growth arises from inductive, reciprocal interactions between mesoderm on the flank of the embryo and its overlying ectoderm. In the chick wing bud the apical ectodermal ridge (AER) of thickened ectodermal cells at the distal tip of the bud is an *organizing center* which, if removed, prevents further development (see Fig. 2.2). The apical epidermal cap in the urodele limb regenerate is normally assumed to be the equivalent of the apical ectodermal ridge in the chick wing bud.

Limb development along the proximo-distal axis (see Fig. 2.3) involves structures associated proximally with the single humerus and distally with multiple digits, as well as asymmetrical development along the anterior-

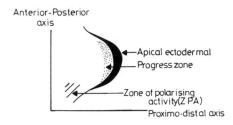

Figure 2.2.   Development zones in early chick wing bud.

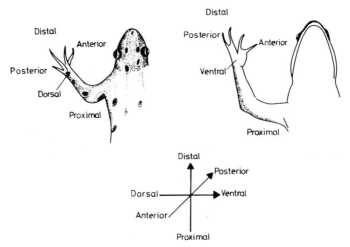

Figure 2.3.   Amphibian forelimb showing major axes.

posterior axis to give right- or left-handed limbs. The dorsoventral axis is determined early in the limb/wing bud and is significantly influenced by ectoderm (MacCabe et al., 1974).

Transplantations of regions of mesoderm have led to the identification of distinct areas within the wing bud, influencing the development and symmetry of the wing (see Ham & Veomett, 1980). Digits normally develop from the postaxial region, that is, posterior to the long axis of the developing limb of the bud. If the posterior region of the bud is transplanted underneath the apical ectodermal ridge at the most distal part of the bud, extra wing parts develop, with digital symmetry identical to that of the host bud. If the posterior region is transplanted to the anterior region, duplicate structures arise, with inverted symmetry. Because of this effect this posterior region has been called the *zone of polarizing activity* (ZPA) (see Fig. 2.2).

A further concept which has been introduced is that of the *progress zone* (see Fig. 2.2) (see Tickle et al., 1975; Tickle, 1981). This is the region of dividing cells adjacent to the tip of the developing limb and thus very close to the apical ectodermal ridge. The dividing cells eventually extend into a region out of the influence of the epidermis. Those cells which leave the zone first provide the most proximal tissues, and the last cells to leave, the most distal tissues.

2.2.3.   Limb Regeneration

Since 1970 a number of suggestions have been advanced to explain positional aspects of limb development or regeneration, normally also including other instances of polarized growth in invertebrates. (For overview of main theories see Tank & Holder, 1981). The introduction of the concept of *positional information* is usually attributed to Wolpert (1969; Wolpert et al., 1975), based on his interpretation of development in *Hydra* and the chick wing bud. A recent outline of the theory (Tickle, 1981) initially distinguishes two stages—the giving of information to the cell dependent on its position, and the transduction of the message to effect the appropriate phenotypic output. Positional information must be represented three-dimensionally; in the wing bud Wolpert proposed that, for the proximo-distal axis, it was related to the time cells spent in the progress zone, close to AER. The shorter the time spent in the progress zone, the more proximal were the structures developed. Information along the anterior-posterior axis was postulated to be related to the distance the cells were from the ZPA—those closest being the most posterior. Determination of the dorsoventral axis is unknown.

The role of the polarizing zone and the operation of the Wolpert model or its derivatives is controversial; other interpretations of the developmental observations continue to be offered. Stocum and Fallon (1982) suggest that the forelimb region of the urodele embryo acquires its transverse (anterior-posterior—AP—and dorsoventral—DV) axes simultaneously, rather than sequentially, by the actions of posterior and dorsal polarizing regions. Determination of these axes is required before proximo-distal growth can occur. The existence of morphogens released by the polarizing regions is inherent in such interpretations and continues to be forthcoming (e.g., MacCabe & Richardson, 1982), although without complete identification.

Another influential theory is the *polar coordinate model* of French and colleagues (1976) and Bryant and colleagues (1981) (see Fig. 2.4). This was derived from work on cockroach leg development and regeneration. As implied in its name, information is conveyed to the cells using radial and circumferential coordinates rather than orthogonal ones. This model makes precise predictions about patterns of development which should occur in surgically constructed amputation/graft procedures. Two rules were formulated: (1) If a part was deleted and the cut surfaces juxtaposed, regeneration occurred following the shortest route of intercalation. (2) Distal transfor-

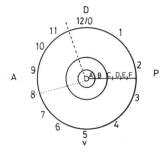

Figure 2.4. Polar coordinate model for positional information (after Bryant & Iten). Reprinted by permission of Academic Press (*Developmental Biology*, Vol. 50, p. 224, 1976).

mation required a complete set of circumferential positional values to be present at the amputation plane. This latter requirement cannot be absolute — a number of workers have constructed amphibian double half limbs which behave in contravention to the "complete circle rule" (Slack & Savage, 1978).

Increasingly sophisticated experiments (Holder et al., 1980; Stocum, 1980) with fore or hind limbs grafted to construct double half-anterior or double half-posterior limbs which were subsequently amputated through the plane of the graft showed regeneration to occur more frequently from fore than from hind limbs and from double posterior rather than double anterior constructs. Abnormal digital symmetry was usually seen.

Recently, approaches have been made toward a chemical interpretation of the positional coordinates. Maden's (1977) positional averaging model led him to a number of postulates:

1. There is a proximo-distal gradient of cell membrane states in the mesodermal tissue of the limb field. The value for the epidermis is zero.

2. Mature tissues do not show this — dedifferentiation must occur before the membrane-coded positional signal can be read.

3. Apposition of the cell displaying its positional value to a nondisplaying cell induces the latter to display and thus dedifferentiate.

4. Cells differing in their positional values by more than a critical amount promote intercalation and the averaging of the values of the cells with their neighbors. When cell division occurs, one cell retains its original value and one intercalates.

5. If there is no positional disparity, cells redifferentiate, but no induction of the positional display occurs.

Throughout the discussion the potential of the cell surface for intercellular communication was stressed.

Extending the interpretation at the cellular level, Stocum (1980), as a result of experiments with axolotl double half limbs, proposed that a blueprint of what was to be regenerated exists at the tip of the amputated limb after epidermal wound healing. Pattern regulation in the blastema is thus already imprinted on the cells in the cross-sectional transverse axes from the parental limb tissue. In normal regeneration, therefore, only the proximo-distal axis has to be restored, which, Stocum suggests, can be achieved by averaging intercalation. Specification of the axes in the stump at the amputation plane arises from the contrast in properties between the cells of the outermost layer of the blastema and the epidermis with which they are in contact.

Confirmation of the early determination of anterior-posterior and dorsoventral axes comes from grafting experiments in *Ambystoma*, using species-specific morphological and cytological markers to distinguish cells of host or donor origin (Stocum, 1982). With this technique for unequivocal assignment for the origin of cells in the regenerate, and in contrast to some earlier reports, 100% of the regenerated structures maintained the handedness of their origins, even with AP or DV axes in the grafts reversed. When host handedness was observed, the cells were always of host origin. The experiments also demonstrated the persistence of positional information through the processes culminating in the establishment of the blastema.

The origins from which the axes are specified are unknown and highly controversial. Bryant and colleagues (1981), in revising their polar coordinate theory to involve only local cellular interactions, stressed the potentialities of connective tissue, which we saw earlier was important in the development of a number of epithelial tissues in culture (Section 6.4.2).

The importance of skin in specifying position in axolotl limb regeneration is indicated from the experiments of Maden and Mustafa (1982) and Slack (1983), in which skin was grafted onto one or another surface of the limb and, after the graft was established, the amputation was performed through the region. A number of points emerged: the skin graft affects pattern regeneration, and (cf. Stocum, 1982, preceding) its capacity to do this is very persistently retained by the grafted cells. Further, information specifying the posterior axis is dominant to that for anterior. In reviewing his own and other evidence, Slack favors the dermis rather than epidermis as the source of positional information.

Slack (1980a) has carried still further attempts to interpret the positional requirements of regeneration biochemically. In his *serial threshold theory of regeneration* he considers the response of the genome to the information signaled into the cell. Because the overall commitment of the cell to a particular morphological endpoint (e.g., epidermis, myotubule, chondrocyte) must be all-or-none, it is usual to envisage cell interactions yielding very sharp boundary conditions. In current biochemical terms these imply allosteric regulation, often with high Hill coefficients, although kinetic solutions are also available (Lewis et al., 1977). Once an effector has achieved these conditions, a gene can be switched on. Positive (serial threshold theory) or negative feedback from the gene product enhances or diminishes the effect (for evidence from cell cultures see Section 5.2.2). Slack (1980a) postulates that, in embryonic development, all relevant switches are on along a particular axis. Distal transformation entails the sequential switching off of increasing numbers of genes in the lineage. Once more, communication between the cells resulting in the reception of the signal at the genome is thought to involve glycosylated surface markers. Glycosyl transferases affecting these residues could be encoded by genes controlled by the switch mechanism (Slack, 1980a).

Attempts are now being made to identify possible positional markers by two-dimensional electrophoresis. No markers have yet been identified, but there are still technical limitations in the analysis (Slack, 1982).

2.2.4. Mathematical Models of Pattern Formulation

Mathematical models for development differ widely in their theoretical approaches and biological premises. In line with the then-current views about gradients and polarity, Turing (1952), considering the classical embryological data available, proposed the *reaction-diffusion mechanism*. These ideas were developed by Gierer and Meinhardt (1972); they assumed the presence of two interacting systems. The first was an activator promoting development, whose synthesis was controlled by feedback with effects that were localized, since a slow rate of diffusion was postulated. The second factor was an inhibitor which diffused rapidly and was effective over a relatively large range (see Goodwin, 1976). The two regulators reacting together and diffusing through a tissue would set up a homogeneous equilibrium which could be distorted by random fluctuations to allow patterns to be formed. In a ring of cells, for example, stationary waves could be set up, giving rise to a pattern of tentacles in *Hydra*. Gierer (1981), extending these conditions in the hydroids, reviewed conditions for pattern formation from two morphogens. He concluded one of the components must be self-enhancing and the other cross-inhibitory, with the inhibitor prominent enough and fast-acting enough to yield a stable field. The range of activation must be below the size of the field where the pattern was exhibited, and less than the range of action of the inhibitor. Similar conclusions were reached by Müller (1982), who further suggested, "The development of patterns in multicellular organisms is governed more by common formal principles than by common molecules or genes."

In recent experiments by MacWilliams (see MacWilliams, 1982) on regeneration in *Hydra,* the zone normally influenced by the activator is the head. After its amputation a new activated zone must be produced in a region previously inhibited. It was predicted that injury caused local production of the activator superimposed on basal, low levels of activator and inhibitor. His results could be reconciled with a proportion-regulatory version of the Gierer-Meinhardt model.

With the discovery of pulsatile signals for *Dictyostelium* aggregation and the existence of metabolic rhythms and their potential entrainment with circadian ones in multicellular organisms, interest arose in the possibility that developmental regulations might involve temporarily variable signals. Theoretical implications of pulsed signals interacting to provide spatial information have been examined by Goodwin and Cohen (1969) and Goodwin (1976). Thom's (1975) *catastrophe theory* envisaged spatial and temporal discontinuities where local conditions diverged sharply from the surrounding environment. These would be sensed by the developing cells to bring about the phenotypic change. Topological constructs were derived from this approach consistent with various structures, but many difficulties arise in equating biological and mathematical parameters in this treatment.

Especially significant is the importance given to diffusion, both in current biological theories and in mathematical treatments. Certain weightings allow reaction-diffusion and catastrophe treatments to be reconciled (Schiffman, 1981). The flexibility and power of the mathematical armory suggests that mathematical descriptions will be available to formalize and probe the implications of whatever biological mechanism for positional information finally emerges.

3

FUNCTIONAL RESTORATION AND COMPENSATORY HYPERPLASIA IN MAMMALIAN SYSTEMS

3.1. SKIN

3.1.1. Skin Repair after Wounding

The ability of skin to be repaired following injury is a familiar demonstration of regenerative ability in humans and one that has received much study. It is also the system from which arose the hypothesis that cell proliferation is normally inhibited by a tissue-specific agent-a chalone (Bullough, 1965).

Classical studies on skin repair were made by Wigglesworth (1937) with the abdominal epidermis of the blood-sucking bug *Rhodnius*. Beneath the cuticle of the adult insect there is a single layer of epidermal cells which normally show no further growth, that is, there are no mitoses. Any mitotic figures after injury must therefore be due to the repair process. After a small incision the cells surrounding a wound enlarge and by 12 h crowd toward the periphery of the wound. The cells in the area from which migration has occurred start to divide two to four days after wounding. When migrated cells have covered the wound, cuticle secretion commences. Wound healing is completed in three to four weeks. Evidence was offered consistent with the release of nonspecific peptide chemotactic agents derived by proteolysis from the injured tissue. These experiments show the capacity of the epidermis by itself to respond to wounding.

The response of mammalian skin to injury (Schilling, 1968; Jennings & Florey, 1970; Bourne, 1981; Robbins et al., 1981) follows very closely that of the urodele limbs to amputation. The process has been studied after various types of injury (see Schilling, 1968) in a number of systems, including rats, hairless mice, and rabbits carrying transparent chambers in their ears. Vascular changes are very fast, vasoconstriction occurring transiently within a few seconds following the release of vasoconstrictors from damaged cells. Arteriolar dilation follows quickly in the area of the wound, the capillary bed opens, postcapillary venules dilate, and vascular stagnation follows, with associated hypoxia and increases in capillary and venule permeability. The site of injury is rapidly covered by a clot, into which leukocytes infiltrate and stick to damaged endothelial surfaces. Platelets similarly adhere, filling in the clot. Surface bonds between contiguous endothelial cells decrease, so fluid leaks out from the vessels and migration of more leukocytes to the site of entry is facilitated. Polymorphonuclear leukocytes are seen first, followed by monocytes and macrophages. The presence of the macrophages heralds the commencement of the healing stage, since these cells will eventually digest fibrin and collagen in the mature clot.

Migration of leukocytes (Zigmond, 1982; Lackie, 1982) is aided by chemotactic factors (see Section 6.5) such as fibrinopeptides released from fibrinogen by thrombin and degradation products from C′3 and C′5 components of complement. Vascular permeability is enhanced by bradykinin, released by peptides activated through the kallikrein cascade (see Robbins et al., 1981). Bradykinin also activates prostaglandin synthesis.

In rats mitoses are evident in the epidermis at the periphery of the wound after about 24 h (see Schmidt, 1968). Adhesion between the epidermal cells is diminished, perhaps because surface components in the newly divided cells have not matured, so the cells can move out over the wound. Mitoses in fibroblasts beneath the epidermal cover increase a day or so after the increase in the epidermis. Thrombin, which may be present in the wound area at least until the clot is resolved, is a powerful mitogen for fibroblasts in culture (Low et al., 1982 and Section 4.2.2). By four to five days new collagen fibers are detectable, showing the reformation of supporting connective tissue. Elastin appears later. Differentiation of the epidermal cells to reform the keratinized layers is thought to recapitulate normal development (Iversen et al., 1968; Potten & Allen, 1975).

Capillary reformation was described when limb regeneration was discussed. Specialized cells in skin, such as hair follicles or sweat glands, are not usually restored.

3.1.2. Hormonal Control of Epidermal Proliferation

The idea that growth might be controlled by the balance between stimulatory and inhibitory factors is an old one (see Weiss & Kavenau, 1957) which received direct support from the experiments of Bullough and Lawrence (1959–1960) on the number and sites of mitoses in mouse pinna epidermis after injury. The mitoses were not restricted to cells not in contact with one another (i.e., relief from contact inhibition was an insufficient explanation), the effects were local, being detectable only 1 mm from the wound, and they were tissue specific—only the epidermis responded. The localization of the response was thought to favor diminution in the concentration of a locally acting inhibitor rather than the release of a positive stimulator from the injury (*wound hormone*), since the latter, it was supposed, would be more widespread in its effects. Bullough and Lawrence suggested that epidermal cells normally produce an inhibitor—a chalone—restricting their own rate of proliferation, perhaps by restricting progress through the G_1 phase, that is, the chalone exerted negative feedback control. Crude preparations of the inhibitor were isolated and reported to be glycoprotein (M_r = 30–40 kDa) and very unstable. Biological assays were required, usually by ^{3}HTdR incorporation into DNA ± chalone. Essential evidence to show that the incorporation was associated with mitoses was also provided.

Confirmation of the existence of tissue- or cell-type-specific inhibitors has been forthcoming (see Iversen et al., 1974; Lenfont et al., 1976; Allen & Smith, 1979), but purification of any chalone has proved remarkably difficult (Allen & Smith, 1979), with the nature of the molecule increasingly uncertain. Recent studies favor lower-molecular-weight compounds (c. 2 kDa), perhaps bound onto glycoproteins, which are of themselves inactive.

Counterbalancing chalones is the idea of cell-type-specific growth promoters. Epidermal growth factor (EGF) (Cohen & Taylor, 1974 and Section 4.2.2), which is present with nerve growth factor in mouse submaxillary glands, is a polypeptide of M_r = 5–6 kDa, which may be associated with proteins having the properties of arginine esterases. They have marked effects on epithelial tissue growth in embryo and on fibroblast growth in culture (Chen et al., 1977; Phillips & Cristofalo, 1981); epidermis derived carcinoma cells may carry EGF receptors (Schlessinger & Geiger, 1981). The importance of EGF in adult epithelial tissue growth is unclear, but secretion of the factor from the periductal cells in the submaxillary glands occurs in vivo in response to α-adrenergic agonists.

As will have been apparent from invertebrate and limb regeneration, local cell interactions are thought to be of major importance in tissue growth and organization. Specific nonprotein growth inhibitors have been isolated from *Hydra*, as have growth-promoting polypeptides. Nevertheless, ideas may be shifting from a single growth-regulatory mechanism for a particular cell type towards a variety of such mechanisms, more than one of which may be effective in a cell at any one time and none of which is essential.

3.2. THE HEMOPOIETIC SYSTEM

Cells within the hemopoietic system are able, even in adult animals, to respond to cell depletion or increased functional demand by increased proliferation. The system comprises cells of the lymphoid system and those of the erythroid and myeloid series from which are derived terminally differentiated erythrocytes, platelets, macrophages, and granulocytes (see Fig. 3.1). Because of the mechanism of differentiation adopted in the lymphoid

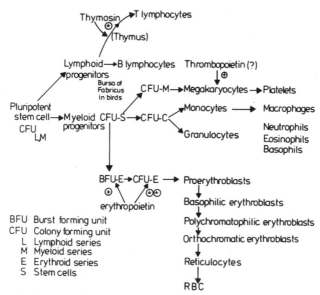

Figure 3.1. Hemopoietic cell lineages (Nienhuis & Benz; Weatherall & Clegg). Reprinted by permission of the *New England Journal of Medicine* (Vol. 297, p. 1319, 1977) and Blackwell Scientific Publications (*Thalassemia Syndromes*, 3d ed., p. 51, 1981).

cells and erythrocytes —rearrangement and/or partial or complete DNA exclusion—redifferentiation is not possible (Section 8.2.3), and probably all the cells are derived from a common pluripotent precursor (see Section 3.2.1). The differentiation pathway followed by the progeny of the common stem cell is frequently determined by local environmental factors. Additionally, cell-line-specific hormonal control over the proliferation and development of erythrocyte progenitors is shown by erythropoietin. Classical physiology further identifies one means of feedback control over erythrocyte proliferation through the effects of variations in the partial pressure of O_2 acting on the kidney. Some aspects of lymphoid tissue proliferation will also be considered.

3.2.1. The Basic Strategy and the Erythroid Series

In the late 1940s and early 1950s the radiosensitivity of hemopoietic tissue was recognized. Rats and mice receiving X-irradiation in the lethal range could be rescued if given bone marrow transplants from closely related donors. If the donated cells were chromosomally distinguishable, it was seen that the entire hemopoietic system (see Fig. 3.1) could be repopulated from the marked cells (see Barnes et al., 1959; Weatherall & Clegg, 1981), indicating a common origin for all the cells of the system. In humans with chronic granulocytic leukemia, carrying the Philadelphia chromosome, the lymphocytes were not so marked, showing that early divergence occurs between the lymphoid and myeloid series (see McCulloch & Till, 1971).

Stem cells (burst- and colony-forming units—BFU and CFU) (see Fig. 3.1) are difficult to distinguish in hemopoietic tissue and are normally present in small numbers. Techniques were therefore required and are now available by which proliferation of particular cell lines can be followed unequivocally. This may be achieved by extending the techniques already mentioned. Heavily irradiated mice, with as much as possible of their own hemopoietic system destroyed, are injected with chromosomally identifiable cells and the development of colonies monitored in spleen or marrow (McCulloch & Till, 1971). Alternatively, long-term bone marrow cultures can now be maintained (Dexter et al., 1977), in which stem cell proliferation and granulopoiesis is shown for two to three months. In these cultures an adherent layer of mixed cell types, including fibroblasts, macrophages, endothelial cells, and fat cells, is essential (Dexter et al., 1981; Fialkow, 1982) (see Section 6.4.2). If anemic-mouse serum is added, or normal mouse serum plus erythropoietin, mature nonnucleated erythrocytes are seen. In the presence of serum from the anemic mice adherence is noted between the developing erythroid cells and monocytes (Dexter et al., 1981).

Stem cells repopulating the spleen (CFU-S) belong to the same clone and are found also to give rise mainly to cells of the erythroid series (Trentin, 1971), whereas, if the donated cells repopulate the marrow, neutrophilic granulocytes predominate. If a plug of marrow stroma is implanted into the spleen, cells in that stroma develop into the granulocytic series, establishing the importance of the hemopoietic inductive microenvironment (HIM).

Colony formation in mouse spleen is regulated by two genetic loci, W on chromosome XVII and S1 on chromosome IV. Their effects are recessive and, if homozygous, usually lethal. With hybrid wild-type crosses W/W^v as donors, CFU-S in marrow werre markedly reduced in irradiated recipients, but if normal cells were given to irradiated W/W^v hosts, colony formation was unaffected. Conversely, $S1/S1^d$ recipients did not support the development in their spleens of normal donated cells, indicating that the S1 locus was involved in the control of the microenvironmental factor(s) necessary in spleens for CFU-S proliferation (McCulloch & Till, 1971; von Melchner & Lieschka, 1981).

In normal mouse femur spatially and functionally distinguishable microenvironments have been reported (Lord et al., 1975). CFU-S increased in number and rate of proliferation from the femoral axis toward the bone surface, while CFU-C showed a peak midway between the two, with no differences in rates of division. Components involved in the microenvironment are being characterized (Gartner & Kaplan, 1980 and Section 6.4.2). Lipid-laden adipocytes may be essential.

A further point emerged from repopulation studies with donor cells carrying the T_6 chromosomal marker. If a 50:50 mixture of T_6^+ and T_6^- marrow cells was donated, all 49 colonies examined after 7 days had only a single differentiated cell line, which in all but 2 of the colonies was either T_6^+ or T_6^-. By 11–12 days 50% of the colonies had developed a second cell line, but the two cell lines were always of the same T_6 character (Trentin, 1971).

When cells have become committed to a particular differentiated pathway under the influence of the local environment, differentiation then proceeds (see Fig. 3.1) for the erythroid series under the influence of an erythroid cell-line-specific hormone, erythropoietin (Jacobson et al., 1957; Erslev, 1971). This is a 60–70 kDa glycoprotein released from the kidney, whose concentration in plasma increases in hypoxia. Bilaterally nephrectomized animals did not respond to hypoxia by elevated erythropoietin levels. The mechanism by which the partial pressure of O_2 in the blood influences plasma erythropoietin levels is still controversial but may involve activating an erythropoeitin precursor in the kidney (Mirand, 1972; Fisher, 1983). Cells in both BFU-E and CFU-E respond to erythropoietin, which also stimulates maturation of nondividing erythroblasts (see Nienhuis & Benz, 1977; Weatherall & Clegg, 1981). It is highly likely that maturation of progeny from the other committed stem cells, CFU-M and CFU-C, is similarly promoted by cell-line-specific factors (see Queensberry & Levitt, 1979). Erythropoietin stimulates RNA synthesis; its possible basis of action will be considered when signals for regeneration are discussed.

Besides erythropoietin a tissue-specific inhibitor of CFU-E proliferation has been described (Kivilaakso & Rytöma, 1971). The same group also reported the presence of a chalone reducing granulocyte proliferation by blocking entry of the cells into S phase (Rytöma & Kiviniemi, 1968; see Allen & Smith, 1979).

3.2.2.   Proliferation in Lymphoid Tissue (see Eisen, 1980)

Consideration of proliferation in lymphoid tissue will be restricted to three topics—the regeneration of B and T lymphocytes following lymphoid tissue depletion, cell interactions required for the proliferation of selected populations of B lymphocytes, and some aspects of intercellular signaling in the lymphoid series.

Evidence has already been presented for a common, pluripotent stem cell precursor for all cells of the hemopoeitic system (CFU-L,M) and for the early divergence of the lymphoid line from the myeloid series (see Fig. 3.1). The two functionally distinct families of lymphocytes, B and T cells, can be distinguished by their surface antigens (e.g., see Figs. 3.2 & 3.3). T cells acquire their properties in part by passage through the thymus; contact with its stroma is essential (see Goldschneider, 1980; Owen & Jenkinson, 1981). At least two populations of cells in the thymus are recognizable: a minor set of cortisone-resistant, long-lived Thy 1^-, H-2^+ cells in the medulla, and cells in the cortex having the opposite properties.

Lymphoid tissue is very sensitive to ionizing radiation (Anderson & Standefer, 1982). Lymphocytes, unlike other cells, show a highly radio-sensitive interphase death—the fate of most small lymphocytes exposed to

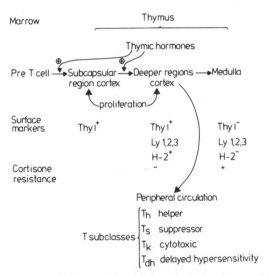

Figure 3.2.   Probable course of T lymphocyte development (see Eisen, 1980; Trainin et al., 1977).

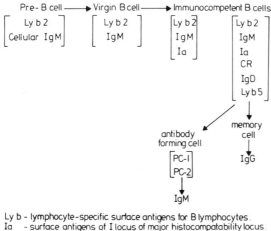

Figure 3.3. Surface markers for B lymphocyte development (see Hämmerling, 1981).

sublethal X-irradiation (see Anderson & Warner, 1976; Quintans & Kaplan, 1978). Lymphocyte repopulation of irradiated animals can relatively easily be achieved by administering lymphoid stem cells. These can be obtained from normal animals by differential centrifugation and the selective removal of contaminating cells by the use of appropriate antisera. Immunological competence is, however, not restored unless the thymus is protected at the time of irradiation or thymic grafts are also given.

The ordered acquisition of differentiated traits is a function of the location of the putative T lymphocytes in the cortical regions of the thymus, reminiscent of the selective differentiation of CFU-S cells into CFU-E and CFU-C in the different regions of the spleen and bone marrow. Few details appear to be available yet about the cell types which are responsible for the inductions.

Additionally, the existence of thymic hormones has long been suggested (see Trainin et al., 1977; White, 1981). Active peptides such as thymosin α_1 have now been isolated, sequenced (Goldstein et al., 1981), and synthesized (Low et al., 1983). It is believed that they act on selected subsets of T cells by increasing intracellular cyclic nucleotides, possibly cyclic guanosine monophosphate (cGMP) (Goldstein et al., 1981). Agents which increase cyclic adenosine monophosphate (cAMP) promote the appearance of surface markers associated with differentiation in cultures of both B cell and thymocyte precursors. B cell surface markers also appear in ordered sequence (see Fig. 3.3). Until the processes affected by the thymic hormone(s) have been fully characterized, it is premature to conclude they play a role in T lymphocyte maturation analogous to the role of erythropoietin in the erythroid series. Nevertheless, it is attractive to concur with Hämmerling (1981), "One can visualize inducible cells . . . which have undergone genomic reprogramming and have implemented the new potential, but which depend for expression on extracellular controls."

The normal stimulus for B cell proliferation is the arrival of an antigen whose structure matches that of an IgM receptor in the membrane of the B lymphocyte, from which a clone of specific antibody-producing B cells will arise. Primary responses require collaboration with macrophages and helper T lymphocytes. For this interaction to occur, certain surface markers must be present in common (Roelants, 1977; Katz, 1979; Unanue, 1981)—antigen receptor sites on B and T cells and Ia gene products of the major histocompatibility complex on macrophages and T cells. The type of antigen recognition on helper T cells is controversial; they may carry an antigen binding site similar to the V_H region of Ig heavy chains (see Tada & Okamura, 1979), but other alternative structures cannot yet be precluded (Jensenius & Williams, 1982).

Electron microscopy has indicated that many helper T cells ($>$ seven) are associated with a single B cell which is in intimate contact with the macrophage (Roelants, 1977). The immunological role of the macrophage (Unanue, 1981) is to present the antigen appropriately to the B cell. Most antigen molecules encountering macrophages are phagacytosed and hydrolyzed, but a small proportion survive, some of which become membrane associated. It is this membrane-bound form of the antigen which is thought to interact with the B lymphocyte.

Consequent to antigen uptake and additional to its processing, macrophages release T cell activating factor(s) (see Rocklin et al., 1980; Ford et al., 1981; Smith & Ruscetti, 1981) now called interleukin-1. This is a globular, 13–15 kDa protein which promotes ^{3}HTdR uptake into cultured T cells. These in turn can release interleukin-2 (15 kDa) (see Taniguchi et al., 1983), which then further stimulates T cell proliferation affecting T_H, T_S, and T_K subclasses.

That T cells also release growth factors by which to influence the B cell with which they are interacting has been demonstrated in elegant experiments with individual B lymphocytes (Pike et al., 1982). Hapten-specific lymphocytes were set up in microculture; proliferation and differentiation were induced in the presence of the specific antigen together with conditioned medium (CM) derived from concanavalin-A (conA) promoted, cloned T cell tumor lines. (With antigen alone, 0.5–0.8% of B cells responded; CM alone, 0.1%; CM + antigen, 5–9%).

T cell dependent antibody formation can be promoted in a subset of B lymphocytes, carrying Lyb-4, by certain types of antigen, such as lipopolysaccharide from *E. coli*, which probably have multiple copies of antigenic determinants/molecule. Stimulation in this case is polyclonal.

The surface changes which antigen binding produces, leading to B cell proliferation, will be considered later (Section 6.1).

The picture which thus emerges from lymphoid cell maturation is of cell-type-specific induction of differentiated pathways, distinguishing the B and T cell lines, followed by the release of a succession of specific factors which promote proliferation of the T cell line, with positive feedback. Primary responses by B cells require interleukins and complex intercellular interactions.

3.2.3. Lymphocyte Cultures

Some types of mammalian lymphocytes, such as those from pig (Forsdyke, 1968a,b), can be maintained in culture and promoted into growth and DNA replication by a variety of mitogens, of which phytohemagglutinin (PHA) and concanavalin-A are very commonly used (Ling & Kay, 1975). Lymphocyte cultures can also be derived from cells of, for example, bovine lymph nodes (Peters, 1975). Cells responding to conA are usually Ly-1$^+$T cells, and those stimulated by PHA are Ly-1$^+$ and Ly-2$^+$T cells. With increasing knowledge of factors required for lymphocyte growth, long-term cultures maintained in serum-free conditions are becoming available (Brown et al., 1983).

Fairly homogenous preparations of lymphocytes can be obtained by differential centrifugation and the biochemical changes following stimulation analyzed (Chapters 7 & 8). The cells are quite well synchronized for the first 24 h after stimulation, that is, into the start of S phase, which is usually detectable at about 21 h. Lymphocyte cultures have been especially informative about early changes in cell membrane properties following growth promotion (Chapter 6).

3.3.   COMPENSATORY HYPERPLASIA IN RODENT LIVERS AFTER PARTIAL HEPATECTOMY

Compensatory hyperplasia—the capacity of liver to be restored after injury—is found in most vertebrates where it has been sought (Harkness, 1957; Bucher & Malt, 1971). In mammalian liver compensatory hyperplasia follows damage from a variety of chemicals such as CCl_4. It is usually studied after partial hepatectomy in rats and mice (Higgins & Anderson, 1931) in which, because of anatomical convenience in their vasculature, the two main lobes can be ligatured off and removed, thus reducing the liver mass by approximately two-thirds (65–68%). The animals recover very easily from the operation, and in 14–28 days the liver mass is restored virtually to that normal for the animal's body weight by growth of the residual small liver lobes.

3.3.1.   Cellular Heterogeneity

The advantages of ease of growth induction, availability of tissue for analysis, and the extensive biochemical knowledge of normal liver are offset by certain inherent difficulties in interpreting the responses observed to partial hepatectomy. First, the cell population is heterogeneous. Some 60–65% of the cells are parenchymal, although these make up about 90% of the cell volume (Bucher & Malt, 1971); the remaining cells include Kupffer cells, bile duct epithelia, and connective tissue cells. Very little is known of the detailed responses of these specialized cells to partial hepatectomy, except that they are slower than for the parenchymal cells (Bucher & Malt, 1971; Widmann & Fahimi, 1975). In a number of cases the behavior of the total regenerating liver mass has been compared with preparations of parenchymal cells obtained from it (see La Brecque & Howard, 1976) with no significant differences emerging, substantiating the conclusion that changes measured in liver preparations after partial hepatectomy reflect the changes in the parenchymal cells.

Even within normal parenchymal cells, however, enzymological heterogeneity is found, with differences in, for example, lactic dehydrogenase and pyruvate kinase activities between cells across the acini from periportal to pericentral regions (Bengtsson et al., 1981). These differences in quantitative expression of different gene products within otherwise similar cells are shown, too, after partial hepatectomy. Parenchymal cells in the periportal regions, although not those immediately adjacent to the vessels, have shorter times before entering mitosis than those in pericentral regions (Rabes et al., 1975). One of the consequences of these differences in times needed before cell division occurs is that parenchymal cells in the periportal regions are likely to divide twice after partial hepatectomy, whereas those in the pericentral region divide only once (see Bucher & Malt, 1971). Further, these differences in response time reduce the synchrony of the preparation, blurring possible sharpness in change of biochemical status.

3.3.2. Other Factors Influencing Rates of Liver Regeneration

Almost any potential variant in the choice of animal or its environment can alter the rate by which the early stages of liver regeneration occurs. Age is important. Neonatal rats show tight synchrony after operation and high mitotic rates of over 30% by 29–30 h. As the animals age, DNA replication is followed less frequently by division, and polyploidy increases. With rats of 200 g body weight it is uncommon to find more than 5% of parenchymal cells in mitosis.

The timing of the first wave of DNA synthesis, which, using ^{3}HTdR uptake, usually commences 15–18 h after partial hepatectomy, is markedly affected by the feeding schedule of the animals (Hopkins et al., 1973) because food uptake in rats is strongly linked with diurnal rhythmicity. The tightest synchrony was obtained with animals on a 12 h light/dark program, with food available for 8 h at the start of the light period and the operation at the end of this period (Barbiroli & Potter, 1971), that is, good synchrony is facilitated by entraining induced growth cycles, diurnal rhythms, and nutrient input.

As with amphibian limb regeneration, ionizing radiation has been used to analyze certain aspects of the response to partial hepatectomy (Holmes, 1956; Ord & Stocken, 1968). If 2–4 Gy total body X-irradiation is given to rats at or up to 6 h after partial hepatectomy, the start of S phase is delayed for 3–6 h. Subsequently cells recover and proceed through S phase, G_2, and into mitosis similar to the behavior of unirradiated animals. Various intermediates accumulate in the residual lobes of the liver as a result of the transient inhibition produced by the irradiation, providing a means by which essential biochemical changes can be analyzed (Chapters 6, 7, & 8). If X-irradiation is not given until the S phase, exposures of $\geqq$ 10 Gy are required to reduce DNA synthesis; the number of subsequent mitoses is then markedly diminished.

3.3.3. Vascular Changes after Partial Hepatectomy

The importance of the portal blood supply has already been indicated from the observations that promotion from the G_0 state of the parenchymal cells occurs more rapidly the closer the cells are to the periportal vessels. The generalization is not absolute, since cells immediately adjacent to the periportal vessels are not the first to go into mitosis. Access to rate-limiting nutrients is probably implicated, as increasing amino acid availability is able, of itself, to promote DNA synthesis (Short et al., 1972).

Immediately after operation the whole of the portal blood is forced to flow through the residual lobes. Potentially this can alter the extent of saturation of hormone receptors or solute transporters (see Ord & Stocken, 1980), but classical experiments on the importance of the changed hemodynamic factors (see Bucher & Malt, 1971) suggest that augmented postoperative blood flow is not critical in the regenerative process.

3.4. RENAL COMPENSATORY HYPERTROPHY OR HYPERPLASIA AFTER UNILATERAL NEPHRECTOMY

Aristotle (330BC[b]) first noted that humans could survive with only one kidney, which, with people under the age of 30, becomes substantially enlarged. Functional restoration after unilateral nephrectomy is, however, much more limited than for liver and is strongly age dependent (Bucher & Malt, 1971; Aschinberg et al., 1978). In neonatal puppies (1–5 days old) which have been unilaterally nephrectomized and in which the blood supply of the second kidney has been reduced by about 50%, homeostasis was effectively maintained, whereas in older litter mates (2 months) glomerular filtration rates were only about half those in control animals (Aschinberg et al., 1978). Even in very young animals new nephrons are not formed, and the glomerular structure shows no sign of regeneration. In the very young animals renal tubular epithelium shows hyperplasia, with increased [3]HTdR uptake into nuclear DNA and mitoses especially in proximal tubules (Davies & Ryan, 1981). In older animals hypertrophy is seen with increases in RNA and protein/cell with little evidence for replication or mitosis.

Complementary to surgical nephrectomy there have been many studies of repair in animal kidneys following injury induced by toxic agents, especially those potentially causing environmental and industrial hazards to humans. Acute renal necrosis is induced by, for example, heavy metals,

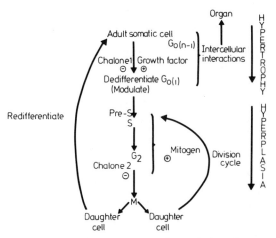

Figure 3.4. Cell fates in compensatory hyperplasia and regeneration.

after which regeneration is seen of proximal tubular cells. If D-serine is administered (Peterson & Carone, 1979), cell death is localized and repair follows along a template provided by the basement membrane (Section 6.4.2).

3.5. SUMMARY (Fig. 3.4)

After the trauma of the injury has been overcome, the first problem faced by systems in metazoan animals capable of regeneration is to detect that tissue depletion has occurred. This is thought to be achieved by:

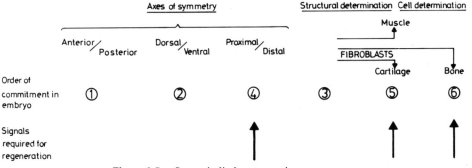

Figure 3.5. Stages in limb regeneration—summary.

1. Loss of feedback signals inhibiting cell proliferation, that is, implying a fall in extracellular chalone concentration.

2. Loss of functional capacity causing a release of growth promoter from some extracellular site, for example, anemia and erythropoietin from kidney.

3. The presence of extracellular stimuli promoting residual growth/proliferation, for example, relatively increased nutrient supply, O_2 stimulation in hydroids, or by combination thereof.

Two types of cellular response follow:

1. Differentiated cells are modulated to a greater (blastema) or lesser (liver) extent and are promoted from a G_0 state through S into division, often increasing in size before DNA synthesis is initiated.

2. Partially differentiated cells (committed progenitors of some cell lineages, e.g., BFU-E) proliferate to give daughter cells, which eventually move into the G_0 state.

Intercellular interactions lead to renewed differentiation and, in a limited number of cases, spatially controlled reformation of complete structures (see Fig. 3.5). In invertebrates total respecification of axes of symmetry, organ structure, and cell type can occur. In amphibian limbs respecification of distal information is required, but AP and DV information is retained from before the amputation. How the spatial information was initially imparted is uncertain, nor is it clear whether, once imprinted, the information is transmissible without continuing specification.

4

THE SIGNALS

For regeneration or compensatory hyperplasia to be achieved, cells must become informed of the need for proliferation, and the processes required for tissue replacement and the restoration of missing structures must be set in train. The great importance currently attributed to intercellular interactions has already been noted. In determining whether and how intercellular signals are involved, we are seriously impeded by our ignorance about nonneural mechanisms used in invertebrates, in which the greatest potential for regeneration is seen. In Amphibia, too, our knowledge is incomplete, so we are overdependent on experiments with mammalian models showing compensatory hyperplasia. Any hypothesis based on these data may well be inapplicable in the quite different circumstances obtaining in poikilotherms and invertebrates. A further uncertainty is the extent to which extracellular regulatory factors used in adult animals with a multiplicity of highly differentiated cells and complementarily specialized membrane receptors are relevant to the blastema. Here there is modulation and respecification of cell properties, very possibly causing the cells temporarily to become more like their embryonic progenitors.

An additional approach to identifying the regulatory factors in regeneration has been the study of growth, proliferation, and differentiation in embryonic cell and tissue cultures. This has yielded a vast quantity of data, among which it can be difficult to discern the results most likely to be significant in vivo.

4.1. THE REQUIRED RESPONSES

For restoration of function and form certain essential changes in the behavior of adult somatic cells are required:

1.   Most differentiated cells are in a G_0 state (Section 1.1.2; Fig. 1.1); reentry into the growth cycle is needed, followed by DNA replication and cell division. Considerable changes have been detected in appearance, membrane properties, and protein synthesis as cells move out of G_0 (Section 1.1.2); this is seen, for example, in the cells beneath the epidermis as the blastema is established in an amputated limb stump. Even after partial hepatectomy where liver function is only minimally perturbed as the residual cells move into G_1, immediate alterations in the properties of parenchymal cells are detectable (Section 7.1).

2. A minimum cell size is needed before DNA replication can be initiated (Section 1.1.2). Often cell growth is required before this minimum size is reached.

3. Where complete regeneration occurs, respecification is needed of the required range of differentiated cells in the appropriate positions. This reprogramming occurs within cells of the same lineage. Cells of ectodermal origin, for example, do not replace those from mesoderm.

4. In cases such as hemopoietic tissue, in which restoration is initiated from stem cells, promotion into an active growth cycle is again obligatory.

Signals reaching regenerating cells can therefore be considered in four classes:

1. Those promoting events necessary for cell growth.
2. Those promoting cell proliferation.
3. Signals specifying a particular differentiated state—osteocyte rather than chondrocyte, B class lymphocyte rather than T class.
4. The positional signals controlling the pattern of cell assemblies.

Many of regulatory factors in the second class with mitogenic activity also, possibly inevitably, promote cell growth, but the reverse does not hold. Increases in cell growth and size do not necessarily lead to DNA synthesis and cell division.

4.1.1. The Nature of the Signals

Because there are a large number of factors which can promote cell growth, it is necessary to restrict our consideration of signals to those agents, released extracellularly, which in low concentration—0.1 nM–1 μM—affect cells where they are bound by specific high-affinity receptors (see Table 4.1). Consequent upon their binding, various means of transduction (adenylate cyclase activation, increased intracellular Ca^{2+}, etc.) are used to affect cell metabolism, structural organization, and/or protein synthesis. Amplification of the effects occurs at one or more stages in the response. The signals may affect one or a small number of cell types only, for example, erythropoietin affecting burst-forming and colony-forming erythroid cells (BFU-E, CFU-E) (Section 3.3.2); some hormones in adult organisms are much less selective. The signals promoting regeneration or compensatory hyperplasia are likely

TABLE 4.1.   AGENTS PROMOTING MITOSIS IN SELECTED MAMMALIAN SYSTEMS

	Liver Hepatocytes	Lymphocytes	Fibroblasts
Peptides			
EGF	+		+
Insulin	+		+
Cell-type-specific peptides	+	+ (Interleukins)	+ (FGF, PDGF)
Lipid-Soluble Growth Promoters			
Prostaglandins	+	+	
Lectins			
Phytohemagglutinin (PHA)		+	
Concanavalin-A (conA)		+	
Cyclic Nucleotides			
Intracellular			
cAMP	+	+	+
Ions			
Ca^{2+}/A23187	+	+	+

to be cell type specific but may require or be facilitated by the concomitant presence of less-specific maintenance hormones affecting growth or metabolism, such as growth hormone or the thyroid hormones.

4.2.   GROWTH FACTORS

4.2.1.   Tissue-Specific Growth Factors

CLASSICAL TROPHIC FACTORS

Apart from vitamins the first growth factors to be intensively studied were the trophic factors from the anterior pituitary. The outcome of this work has been very influential in current thinking about less-well-defined growth factors involved in development and regeneration. It may therefore be useful to review some of the salient experimental approaches employed in earlier studies, to indicate what may be needed if greater understanding is to be obtained of the role of humoral factors in cell interactions in growth and differentiation. Adrenocorticotropin (ACTH) will be taken as the example, since information about it is well advanced.

Isolation and Characterization. Classical endocrinological experiments on the effects of extirpation and its reversal by administration of extracts from the anterior pituitary established that this gland was required for the maintenance and function of the adrenal cortex. In the late 1940s and early 1950s kilogram quantities of animal pituitaries were collected and used by Li and his colleagues to isolate pure ACTH, to determine its amino acid sequence and identify the residues essential for its activity. Recent work has shown the relation of ACTH to a number of other pituitary peptides, indicating that it, lipotropins, melanophore-stimulating hormone (MSH), and β-endorphin may be derived from a common precursor, proopiomelan-ocortin (see Herbert & Uhler, 1982). Different processing of pre-mRNA transcripts in different cells at different stages in development could cause the production of varying amounts of regulatory factors. Additionally, or alternatively, the production of multiple peptides from a single precursor offers a means of coordinated synthesis of functionally related products (Herbert & Uhler, 1982) (Section 8.2.3).

Purification of any hormone necessitates sensitive and specific assay procedures usually operating in the range 0.1–10 nM. The trophic hormones of the pituitary affect target glands, whose products are exceptionally active physiologically. Chemical characterization and detection of steroids synthesized in the adrenal cortex in response to ACTH developed very rapidly in the 1940s and 1950s; radioisotopic techniques allowed identification of the stage in steroid biosynthesis—loss of the cholesterol side chain to give pregnenolone—which was specifically promoted by ACTH. A fall in adrenal ascorbic acid was another early empirical test for ACTH activity in the whole animal.

Techniques now available should obviate some of the problems attendant on obtaining adequate material for characterization of the growth factor. If a peptide growth factor is thought to be produced by particular cells, and its approximate size is known from polyacrylamide gel electrophoresis, isolation of its mRNA is theoretically possible, opening the door both to the determination of the amino acid sequence and to cloning.

One of the problems in the identification of a compound as a mitogen is that mitosis itself is not a very convenient assay. The process is incompletely understood; there are many possible rate-limiting stages, some of which can be stimulated nonspecifically, and there is often a time lag of at least 12 h before increased numbers of mitoses are visible.

ACTH Receptors.   Identification of receptors is usually approached by first making a hormone of very high specific radioactivity. Lactoperoxidase catalyzed iodination is commonly employed but may inactivate the hormone— a problem with ACTH (Schulster & Schwyzer, 1980).

Peptide hormones normally bind onto protein or glycoprotein on the plasma membrane. Internalization of the ligand-receptor complex may follow (see following for EGF and insulin), but frequently much of the activity of the peptide is exerted when it is bound onto the surface. Experiments have purported to demonstrate this by linking hormones to microbeads or inert macromolecules, such as cellulose, which cannot enter cells but still give stimulation. Unfortunately it is impossible to prove that dissociation has not occurred of the hormone from its carrier.

Various preparations can be employed for the binding studies, but because techniques for plasma membrane isolation may cause ligand-receptor dissociation, whole cell preparations or cultures are often used. With the former, methods employed to disperse the cells can alter hormonal sensitivity. Cells from the different regions of the adrenal cortex have been separated and their response to ACTH examined (Tait et al., 1980). With unseparated cells 3000 high-affinity sites/cell (K_D^a = 0.25 nM) and 30,000 low-affinity sites (K_D^b = 10 nM) have been reported (Schulster & Schwyzer, 1980).

Once high-affinity membrane receptors have been identified, the potential arises for their isolation and the raising of monoclonal antibodies (Milstein, 1981) which specifically bind to them and so block the action of the normal hormone. When confirmation of growth stimulation is needed, or when the number of cell types sensitive to stimulation is uncertain, the availability of antibodies against growth-factor-specific receptor is enormously helpful.

Cyclic AMP as Secondary Messenger for ACTH.   Mediation of the effects of ACTH on the adrenal cortex was one of the first examples offered for the role of cAMP as secondary messenger (Grahame-Smith et al., 1967). Four questions have been especially persistent and will be paralleled in later discussions on the mechanisms of transduction of other growth-promoting agents: (1) the role of Ca^{2+} in the stimulation (Berridge, 1975); (2) the proportionality between cAMP release and steroidogenesis (Tait et al.,

1980); (3) how protein phosphorylation promotes cholesterol conversion to pregnenolone; (4) whether all the effects of ACTH, especially the need for protein synthesis and the growth of cortical tissue, can be attributed solely to increased cAMP $\pm$ Ca^{2+} (Halkerston, 1975; Schimmer, 1980).

A role for Ca^{2+} in hormonal stimulation is now an expectation. Its actions as a secondary messenger will be discussed (Section 5.1.2). So will be the possible means by which it enters cells, as part of the pleiotropic response to many hormone-receptor interactions, or is liberated as free Ca^{2+} from intracellular binding sites.

Control of the conditions of incubation has allowed good quantitative correlations between levels of ACTH (0.1–1.0 nM) used to stimulate adrenal cortical cells, the amounts of cAMP released, and the activation of protein kinase and steroidogenesis (see Tait et al., 1980). It also seems likely that maximum promotion of steroidogenesis occurs at cAMP levels which are significantly below those maximally attainable (see Schulster & Schwyzer, 1980). While ACTH is currently believed to promote the increased association of mitochondrial cholesterol with the cytochrome P_{450} oxygenase system, it is not clear how protein phosphorylation assists this (Schimmer, 1980).

Two types of coupling between hormone receptors and adenylate cyclase have been proposed in the past: direct coupling, where the receptor was permanently associated with adenylate kinase, or some type of collision coupling model, where the ligand-receptor complex interacted transiently with the catalytic site (see Braun et al., 1982 and Fig. 4.1). With the identification of the guanine nucleotide regulatory protein as an essential component in the system, a general, unifying, three-domain model has been proposed (Stadel et al., 1982).

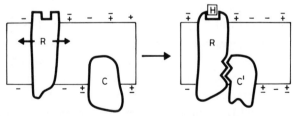

Figure 4.1. Mobile receptor model for hormone receptor coupling. Hormone receptor, R; hormone, H; transducing element, C; activated transducing element, C' (from Houslay & Stanley). Reprinted by permission of John Wiley & Sons (*Dynamics of Biological Membranes*, p. 319, 1982).

Feedback Control of ACTH Secretion.   Feedback control of ACTH secretion by the pituitary was proposed in the 1950s by C. N. H. Long. Alterations in cortical steroid usage would be reflected in their circulating levels, sensed by the pituitary cells, and the levels of synthesis adjusted. High-affinity glucocorticoid receptors have been found in mouse pituitary tumor cells (Watanabe et al., 1973), and if adrenalectomized rats were given dexamethasone, but not if given progesterone or aldosterone, ACTH mRNA levels in their pituitaries decreased (Nakanishi et al., 1977). Further clear evidence that glucocorticoids decrease the contents of proopiomelanocortin in the anterior pituitary by decreasing its mRNA production has been provided by Roberts and colleagues (1982).

4.2.2.   Growth-Promoting Factors

Humoral agents which may be involved in regeneration will next be considered, especially those implicated for liver, lymphocytes, or fibroblast cultures (Gospodarowicz & Moran, 1976; Rozengurt, 1980).

EPIDERMAL GROWTH FACTOR (EGF)

EGF was detected in extracts from mouse submaxillary glands by Cohen and found to stimulate epidermal growth and keratinization (Carpenter & Cohen, 1979). It is a heat-stable polypeptide which has been sequenced (53 amino acids, M_r c. 6 kDa) and which can promote DNA synthesis in a fairly wide range of cells including chick embryo epidermis, human and mouse (3T3) fibroblasts, and regenerating rat liver (Rubin et al., 1982; Schlessinger et al., 1983). In cell cultures exposure to EGF was needed for up to 8 h to cause increased ^{3}HTdR incorporation into DNA by 15–22 h (see Carpenter & Cohen, 1979). Usually the presence of serum was also required, but with primary cultures of rat hepatocytes (Friedman et al., 1981; McGowan et al., 1981) and some other cell types (see Rozengurt, 1980, 1982) serum could be omitted if insulin and cAMP-releasing agents were also present.

The effects of EGF on cell membranes are very rapid (Rubin et al., 1982). With human epidermal carcinoma cells extensive membrane ruffling and filopodial extensions were seen within 5 min (Schlessinger & Geiger, 1981). In rat liver EGF binds onto a membrane receptor (150–180 kDa) (Rubin et al., 1982) with a dissociation constant of 0.2–0.4 nM (Gospodarowicz & Moran, 1976). A tyrosine residue was phosphorylated on a 170 kDa protein which copurified with the EGF membrane receptor (Cohen et al., 1982; Rubin et al., 1982). The extent of phosphorylation was proportional to the levels of EGF used. Inhibition of the phosphorylation by lectins suggests that the receptor is a glycoprotein (Carpenter & Cohen, 1979). Further work has indicated there are two domains: a phosphate receptor site localized on a 45 kDa region accessible from the cytosol, and a region projecting onto the cell surface on which EGF is bound (Fox et al., 1982).

The significance of tyrosine phosphorylation (see also PDGF and insulin) and the autophosphorylation of the hormone receptor are under active examination. In cells transformed by oncogenic viruses, in which tyrosine phosphorylation is greatly enhanced, vinculin—a structural protein involved in cell contacts (Section 6.4.1)— and isozymes of three glycolytic enzymes — enolase, phosphoglycerate mutase, and lactic dehydrogenase—have so far been identified as substrates for tyrosine protein kinase in vivo (Cooper et al., 1983). In cells promoted by EGF and platelet derived growth factor (PDGF), $^{32}P_i$ incorporation into glycolytic enzymes has not been detected, and the range of phosphorylated proteins is still being explored. Phosphorylation into serine may or may not be promoted at the same time. The relevance, if any, of the phosphorylations to cell proliferation and oncogenesis is not yet known (Racker, 1983).

After binding, the EGF receptor complex appears to be internalized. Where internalization has been studied in detail (Pearse & Bretscher, 1981; Besterman & Low, 1983), receptor mediated endocytosis has been found to occur from vesicles which are coated with clathrin (180 kDa, nonglycosylated, highly conserved protein) (Ungewickell & Branton, 1981) on their cytoplasmic surfaces. After internalization the vesicles are thought to lose their clathrin coats and fuse with one another. Alternatively, after the ligand-receptor complexes have clustered into coated pits, the receptor-bound ligand may be transferred to uncoated vesicles, the clathrin remaining attached to the plasma membrane (Willingham & Pastan, 1980). The subsequent fate of the ligand may be determined by its pattern of glycosylation; for EGF substantial breakdown occurs in the lysosomes (see Das, 1982), the receptors being recycled to the plasma membrane. Such a process allows a reduction in the availability of receptors if the concentration of hormone is raised (down-regulation). This may occur by 36 h after partial hepatectomy when EGF-dependent tyrosine phosphorylation is diminished and there is a 67% decrease in the number of EGF receptors (Rubin et al., 1982).

Fibroblast Growth Factors (FGF)

Fibroblast-specific growth factors have been detected in media in which mouse 3T3 cells are growing. The 3T3 cells are especially sensitive to cell density changes, and inhibition of growth at confluence can be reproducibly overcome by serum or derived factors (see Rozengurt, 1980). The early biochemical changes in 3T3 cells after growth promotion are like those in hepatocytes and lymphocytes (Chapters 6–8).

FGF has a primary structure resembling residues 55–143 of myelin basic protein. It, or structures similar to it, could be one of the growth factors released from neural tissue, which we have already seen to be essential for blastemal development in amputated limb stumps in Amphibia. Injecting denervated stumps with FGF stimulated mitotic activity to levels found in normally enervated limb stumps. With mammalian chondrocyte and myoblast cultures both EGF and FGF were mitogenic, but only FGF promoted myoblasts, an effect which was potentiated by dexamethasone to levels identical to those found with fetal calf serum (Gospodarowicz & Mescher, 1980).

Platelet-Derived Growth Factors (PDGF)

PDGF is a polypeptide of about 30 kDa, present in granules in platelets from which it is released when the platelets come into contact with damaged tissue, as might occur prior to regeneration (see Stiles et al., 1981; Stiles, 1983). In contrast to the 6–8 h needed to establish the mitogenic effects of EGF, relatively brief exposure to PDGF promotes growth and DNA synthesis in quiescent fibroblasts. The polypeptide binds onto high-affinity receptors on the cell surface, which get internalized (affinity constant of approximately 1 nM; 3×10^5 receptors/human foreskin fibroblast [Ek et al., 1982]). As with other growth factors we have been considering, PDGF produces rapid changes in membrane transport. Like EGF it stimulates phosphorylation of tyrosine. It is probable that its receptor, like that for EGF, is a tyrosine protein kinase, but it is uncertain whether the range and extent of proteins phosphorylated is coincident with the phosphorylation produced by EGF (Cooper et al., 1982). In 3T3 cells preliminary treatment with PDGF reduces the number of sites available to bind EGF (Leof et al., 1982).

Liver-Specific Factors

The existence of humoral factors specifically promoting liver cell growth and proliferation has been sought since the introduction of the simple technique for partial hepatectomy in rodents (Higgins & Anderson, 1931). Probably the most convincing experiment was that performed in the 1950s, when three young rats were joined laterally along their flanks shortly after weaning, to create trionts. Cross-circulation was established, after which the outer animals were partially hepatectomized. Unequivocally increased DNA synthesis and mitoses were induced in the liver, but not other organs, of the middle rat. Similar results have been obtained in liver autografts after partial hepatectomy (Virolainen, 1964; Leong et al., 1964). The experiments could not determine whether the response was due to the release of humoral factors, stimulating liver cell proliferation, or to the removal of inhibitory factors normally produced by the liver, consequent on the loss of organ mass. Attempts to clarify the interpretation have been continuous, principally by comparing effects of serum from control and partially hepatectomized rats on a variety of liver cell cultures, primary hepatocyte cultures being the most favored preparation. One outcome of this has been the demonstration that many of the hormones normally present in serum from adult animals affect liver regeneration. This will be discussed later; here only possible tissue-specific factors will be considered.

In the 1970s at least three groups of workers produced evidence that liver cell growth could be stimulated specifically by peptide factors. Pickart and Thaler (1973) found a tripeptide containing glycine (Gly), lysine (Lys), and histidine (His) which increased RNA, protein, and DNA synthesis in HTC cells grown on minimal (1%) serum levels. If amounts of serum were increased to those more usually employed, the effects of the tripeptide were no longer detectable. Chemically synthesized Gly-Lys-His or Gly-His-Lys were active, but the equivalent amounts of the free amino acids were not. Morley and Kingdon (1973) showed slight stimulation of [3]HTdR uptake by hepatocytes with a heat-stable protein (26 kDa) from blood of partially hepatectomized rats 12–24 h after operation (see La Brecque, 1982). Further work, with fetal calf serum as source (Morley, 1974), indicated the presence of two possible components, one promoting DNA synthesis and the other ornithine decarboxylase induction in vivo in normal mouse liver at 6 h after administration. The latter effect was dramatic but difficult to interpret because of the well-established lability of ornithine decarboxylase levels (Section 7.2.3).

Further work by Leffert (1974a,b; Koch & Leffert, 1974) used dialyzed fetal calf serum as a source of growth-promoting activity and primary fetal rat hepatocyte monolayers as a test system. A protein of 120 kDa was detected, which stimulated the initiation of DNA synthesis and mitosis, specifically in liver. The presence of such a factor was also detected in serum-deficient conditioned medium obtained from the cultures. The authors concluded that complex interactions among serum factors were involved in the initiation of DNA synthesis. The specialized and hormonally regulated functions of the liver in normal nitrogen metabolism, and the problems in forecasting fates of individual amino acids or peptides in a tissue where compartmentation with respect to amino acid metabolism is frequently invoked, make analysis and extension of this data very difficult.

Exactly the same troubles attend complementary work on the existence of liver-specific growth-inhibitory factors (chalones) (Verly et al., 1971; Sekas et al., 1979; McMahon et al., 1982), which inhibit DNA synthesis but whose relevance to physiological controls operating during liver regeneration is uncertain.

OTHER CELL-TYPE-SPECIFIC GROWTH REGULATORS

Besides the growth factors already mentioned, many cells grown in culture release products in their media which positively or negatively modify in the same or other cell lines (see Nilsen-Hamilton et al., 1980). One such inhibitory protein, from African green monkey epithelial cells (BSC-1) (Holley et al., 1980), has been extensively purified. It is selective in its effects on cells, mouse 3T3 and human skin fibroblasts being relatively insensitive and BSC-1 cells the most affected. Growth inhibition produced by the protein is counteracted by EGF or serum, but the inhibiting protein does not appear to be blocking EGF interaction with the EGF receptor.

Other increasingly well-defined regulators are *nexins*, which are released from fibroblast derived cells in culture. Nexin-1 is a protein which binds covalently to thrombin through the serine in the latter's catalytic site. Binding to nexin-1 prevents the proteolytic action of the thrombin, which appears to be essential for its mitogenicity to fibroblasts (Low et al., 1982).

These examples of growth inhibitors, which are more stable and so better characterized than chalones, offer the best evidence so far for specific regulatory substances being released from cells with the potential to limit their own growth.

4.2.3. Relatively Nonspecific Peptide Growth Factors
(see Tata, 1980; Rothstein, 1982)

INSULIN

The anabolic action of insulin received particular attention in the 1940s from F. G. Young, who suggested the growth-promoting activities of the hormone might be attributable to its capacity, in cooperation with other hormones affecting metabolic homeostasis, to facilitate the effective utilization of glucose and so allow amino acids to be used for protein synthesis. The metabolic functions which are most obviously affected by insulin are easily seen in adipocytes; insulin stimulates transport and metabolism of glucose, inhibits lipolysis, stimulates leucine incorporation into protein, and stimulates pyruvate dehydrogenase, acetyl CoA carboxylase, and glycogen synthetase (see Kahn et al., 1981).

The mechanisms by which these changes are brought about are still very controversial. Probably the commonest hypothesis is that most, possibly all, of the metabolic effects of insulin are due to an alteration in phosphorylation status of key proteins (Larner et al., 1982), perhaps via activation of protein phosphatase-1 (Cohen, 1980). Whether the actions of insulin follow from the release of a secondary messenger from the plasma membrane after hormone binding is still under extremely active investigation (Larner et al., 1982). Obvious complexities arise because of interactions with cyclic nucleotide—and Ca-calmodulin dependent phosphorylations. A well-known and much-studied example is insulin-augmented phosphorylation of the plasma membrane—low-K_m cAMP phosphodiesterase, increasing its activity and so lowering levels of cAMP (see Marchmont & Houslay, 1981).

The most direct way in which insulin could affect phosphorylation would be that its receptor was also a protein kinase, like that for EGF. Support for this view is emerging (see Petruzelli et al., 1982). Roth and Cassell (1983) have shown that the β-subunit of the insulin receptor (95 kDa) is a protein kinase, the α-subunit being the insulin-binding portion, and this has been confirmed by Shia and Pilch (1983), who have shown phosphorylation of tyrosine in the β-subunit of human placental insulin receptor. Modulation of the activity of the receptor by tyrosine phosphorylation, auto-regulated by insulin, has been suggested (Kasuga et al., 1982; Rosen et al., 1983).

Protein and RNA synthesis in many tissues are also affected by insulin, though more slowly (h cf. min) and at higher concentrations (0.1 μM cf. 10 pM) than with its metabolic action (Kahn et al., 1981). Effects on growth are especially evident in cell cultures; partial replacement of serum by insulin allows growth and proliferation to be maintained. With fibroblasts the effects of insulin can be additive, or even synergistic, with other growth factors such as EGF or FGF (Rozengurt, 1980).

Besides insulin there are other polypeptides isolated from serum which have insulinlike effects on growth and metabolism (see Gospodarowicz &

Moran, 1976; Rothstein, 1982). These include somatomedins A, B (IGF-1, IGF-2), and C, which are released from liver in response to growth-hormone stimulation and multiplication-stimulating activity (MSA) in the medium from Coon rat liver cells (which do not require serum supplementation for growth). These insulinlike growth factors (IGF) have some sequence homologies with insulin but usually show relatively greater growth promotion and less evident metabolic stimulation.

Much attention is directed at present to the identity or otherwise of the high-affinity receptors for insulin and IGF. Divalent antiinsulin receptor antibodies can be obtained which, after being bound onto the cell membranes, provoke metabolic responses identical to those following insulin binding to the receptors but which do not stimulate ^{3}HTdR uptake into fibroblast DNA (Kahn et al., 1981). If monovalent F_{ab} fragments of these antibodies are used, insulin receptors are again blocked, but this time without eliciting the metabolic responses. MSA binding was virtually unimpaired and enhanced ^{3}HTdR incorporation into DNA. From these (Kahn et al., 1981) and parallel observations with other systems (Jonas et al., 1982) it was concluded that two receptors are involved: one preferentially binding to insulin and so causing its metabolic actions, and the other preferring IGF-2 and leading to growth. Cross-reactivity between insulin and IGF could occur.

Further work (M. Czech, private communication) has indicated considerable similarities between the insulin receptor and that for IGF-1. The receptor for IGF-2 is different. It does not bind insulin and its stimulation produces neither the metabolic effects nor the facilitation of transport seen with insulin and IGF-1. In cells such as adipocytes, which have both insulin receptors and those for IGF-2, IGF-2 binding onto its receptor is promoted by insulin (Oppenheimer et al., 1983).

The existence of a receptor for the insulin class of growth factors in which ligand binding stimulates growth raises new problems (see Czech, 1982). What mechanism of transduction is used, and how is *growth* promoted? The idea that insulin affects protein synthesis on the ribosomes emerged principally from experiments from Wool and his group (see Wool, 1979), initially showing that incorporation into protein of amino acids from the intracellular pool was stimulated by insulin. It was subsequently established (see Leader, 1980; Thomas et al., 1982; Martin-Perez & Thomas, 1983) that ribosomal proteins could be phosphorylated, that in the 40 S ribosomal subunit the prominently phosphorylated protein was S6, and that enhanced phosphorylation of this protein could be observed in certain rapidly growing cell cultures and regenerating rat liver (see Leader, 1980; Nishimura & Deuel, 1983). In spite of this, correlations between active protein synthesis and ribosomal protein phosphorylation are imperfect (Leader, 1980; Gordon et al., 1982; Nilsen-Hamilton et al., 1982), and the effects of phosphorylation of structural proteins on the structure and function of the ribosome are uncertain. It is noteworthy, though, that ribosomes carrying the most highly phosphorylated S6 are those selectively present in polysomes (Thomas et al., 1982). The cAMP-dependent kinase can promote ribosomal protein phosphorylation in vitro, as does a Ca^{2+}-dependent kinase. Other kinases may also phosphorylate ribosomal protein S6 (Martin-Perez & Thomas, 1983).

Insulin can be internalized and bound onto a number of intracellular sites (Shafie & Hilf, 1981; Goldfine, 1981). Its receptor may subsequently be recycled to the cell surface (Fehlman et al., 1982). Down-regulation of its receptors has also been reported (Goldfine, 1981). It is debatable if the hormone or its degradation products exert any activity through such intra-cellular binding.

One of the problems with which we must be concerned is the possibility that receptors for growth factors change either in number or ligand-binding properties during development or dedifferentiation. Insulin receptors in liver plasma membranes from 15–18 week old human fetuses had approx-imately half the number of receptor sites/100 μg membrane protein than were present after 19–25 weeks. The affinity constant for insulin rose towards birth in both human and rat hepatocytes (Blazquez et al., 1976; Neuffeld et al., 1980; Handlogten & Kilberg, 1982). The authors were satisfied that the changes were not due to alterations in cell population in the liver through this time. Human monocytes in cord blood at birth had considerably greater capacity to bind insulin than those in blood in older children or adults (Thorssen & Hintz, 1977). The numbers of insulin receptors rose in fibroblasts and lymphocytes if high cAMP levels were induced (Thomopoulos et al., 1977) and in noncytotoxic T cell clones during mitogen-induced DNA syn-thesis (Braciale et al., 1982). In fetal rat hepatocytes internalization of ^{125}I-insulin-binding sites was more evident than in adult cells (Autuori et al., 1981).

Effects of Insulin on Compensatory Hyperplasia in Liver. Especially valuable studies of the effects of insulin in regeneration or compensatory hyperplasia have been made in liver. Many early workers (see Bucher &

Malt, 1971; Sigel, 1972) had shown that diverting the portal blood outflow to the vena cava, thus bypassing its normal entry to the liver, delayed or reduced the restoration of liver weight. In a particularly important series of experiments with eviscerated rats, Bucher and Swaffield (1976) established that regeneration in these partially hepatectomized animals occurred in the absence of insulin but was considerably delayed. This delay was diminished by giving insulin, and if insulin and glucagon were administered together, the normal extent and timing of DNA synthesis was substantially restored. This was still seen if the hormones were not given until 6–7 h after operation. On this latter point the authors concluded, "Additional hormones or humoral factors may be involved in the initial activation of hepatic regeneration." In noneviscerated rats insulin levels in blood from the portal vein fell to near zero in the first few hours of regeneration. Further experiments from the same group with insulin antiserum showed it prevented the first wave of DNA synthesis after partial hepatectomy. Increased portal glucagon levels were detected 12 h after operation (Cornell, 1981).

These and other experiments with endocrinectomized rats (see Harkness, 1957) make it clear that restoration of liver mass after partial hepatectomy will occur in the absence of most hormones but much more slowly than in normal animals. With primary adult rat hepatocyte cultures (Leffert & Koch, 1980), EGF, insulin, and glucagon directly stimulated proliferation, unlike calcitonin, parathormone, glucocorticoids, and iodothyronine, which might, however, be implicated in vivo. EGF was effective early in G_1 phase (0–3 h) and insulin/glucagon at 3–12 h.

4.3. LIPID-SOLUBLE GROWTH PROMOTERS

4.3.1. Prostaglandins (see Stryer, 1981; Samuelsson et al., 1978)

Prostaglandins are released from cells in response to many agents, especially those such as thrombin, bradykinin, histamine, and serotonin which are liberated at sites of injury and cause hydrolysis of membrane phospholipids, yielding arachidonic acid from which the prostaglandins are synthesized (see Hassid, 1982). Aspirin and indomethacin inhibit prostaglandin endoperoxide synthetase by blocking cyclooxygenase. Inhibition of some biological function by indomethacin often indicates prostaglandin involvement.

The unstable endoperoxide thromboxane A_2 produces local smooth-muscle contractions; the more stable prostaglandins, especially the PGE series, have specific high-affinity receptors on many cell types and, in contrast to PGF, at 0.1 nM–10 μM activate adenylate cyclase and thus increase intracellular cAMP (see Samuelsson et al., 1978; Zalin, 1979).

Because prostaglandins and thromboxane are likely to be released as a result of the injuries associated with partial hepatectomy, Whitfield and his colleagues (Andreis et al., 1981; Rixon & Whitfield, 1982) looked for possible involvement of these factors in liver regeneration. Indomethacin (5 mg/kg) was found to inhibit DNA synthesis after partial hepatectomy. Further, in calcium deprived hepatocyte cultures, arachidonic acid was very effective in stimulating DNA synthesis—an effect again prevented by indomethacin.

Amounts of prostaglandins released are a function of cell growth rates. Arachidonic acid stimulation of prostaglandin release by 3T3 cells was reduced as the cells became confluent (Hyman et al., 1982).

4.3.2. Steroids

While prostaglandins are strong candidates for participation in the immediate, local, tissue response to cell depletion or limb ablation, steroid hormones are more likely to be involved in the sustained process of repair. Glucocorticoids have been implicated because of their participation in enzyme induction in a wide range of cells. The concept is emerging that the different steroids promote synthesis of selected groups of proteins by binding to specific protein receptors in the cytoplasm, moving as a complex to the nucleus, where the receptor interacts with specific DNA sequences adjacent to the structural genes whose transcription is thereby enhanced (see Edelman, 1975; Payvar et al., 1981; Tata, 1982, and Section 8.2.1). In hepatocytes the synthesis of a number of proteins is promoted by glucocorticoids, dexamethasone being the steroid agonist usually tested.

The physiological effects of the glucocorticoids in intact animals are to enable protein metabolism to be appropriately adjusted to the amino acid and carbohydrate intake. In liver, for example, tyrosine aminotransferase (EC 2.6.1.5) and tryptophanoxygenase (EC 1.13.1.12) are transiently induced after a protein-rich meal, and increased protein breakdown can also be induced in stationary hepatocyte monolayers (Hopgood et al., 1981). In muscle glucocorticoids facilitate the release of alanine and glutamine into the circulation during starvation, probably through the induction of proteases. In rat liver, 17α-hydroxyprogesterone blocks binding of cortisol and dexamethasone to steroid receptors. It also blocks enzyme induction by dexamethasone and the induction of ornithine decarboxylase normally seen 4– 6 h after partial hepatectomy, suggesting cortical steroids may facilitate the early stages in liver regeneration (Thrower & Ord, 1974).

In lymphocytes glucocorticoids induce cytolysis, possibly through protease induction (Macdonald & Cidlowski, 1981). In other cell cultures in which glucocorticoids can be growth inhibitory, a restriction point (see Section 1.1.2) in mid G_1 phase has been detected, before which dexamethasone prolonged G_1 and after which the cell cycle was insensitive. No detectable alterations in general protein metabolism were found (Bakke & Rønning, 1982). Glucocorticoids are involved in regulating genes which may be periodically expressed. They probably promote the synthesis of mRNAs for two sets of proteins, the second of which nullifies (usually hydrolyzes) the first group. The observed consequences of the steroid therefore depend on which effect is dominating at the time of the examination. Stimulation of genes for hydrolases will generally be growth inhibitory.

4.4. THYROID HORMONES

Thyroid hormones have such well-established actions on development that their influence in compensatory hyperplasia and regeneration in vertebrates is not unexpected (Liversage & Brandes, 1982), like the other hormones which influence growth relatively nonspecifically. Recent work with liver and pituitary-derived cells, which synthesize α_{2u} -globulin and growth hormone, respectively, in response to triiodothyronine, has suggested that triiodothyronine binds to a chromatin associated receptor in cell nuclei, regulating a transcriptional or posttranscriptional event which is rate limiting for mRNA accumulation (see Samuels et al., 1982). The receptor appears to bind to a subset of nucleosomes which are preferentially released from nuclei as mononucleosomes by micrococcal nuclease, behavior usually associated with nucleosomes from transcriptionally active chromatin (see Section 8.1.3).

Lee and colleagues (1968) showed that triiodothyronine administration induced $^{32}P_i$ uptake into rat liver DNA after 24 h. The mitotic index also rose. Extension of these experiments by Lieberman and his colleagues (Short et al., 1972) demonstrated that, if rats were infused with amino acids, glucagon, triiodothyronine, and heparin for 3 h, increased [3]HTdR labelling in DNA was detected at 14 h, followed by an increased number of mitoses. The presence of triiodothyronine considerably enhanced the response.

4.5. NUTRITIONAL REQUIREMENTS FOR RESPONSE

Although nutrients are not contained within the definition of *signal*, there is excellent evidence that, for cells in culture and even for some in vivo, the presence of appropriate nutrients, especially amino acids, may be strongly growth promoting and, for liver, may produce increased DNA synthesis. The importance of metabolic controls for cell cultures has long been recognized (Eagle, 1965; Holley, 1975). In intact animals it is unlikely that circulatory levels of amino acids fall below 0.01–0.04 mM, which Eagle estimated to be the minimum concentration required for cell growth, but fluctuations in the levels of less-abundant amino acids, such as tryptophan and methionine, may be associated with periodic fluctuations in protein synthesis in liver (Munro, 1969).

Administering casein or its equivalent amino acids to intact rats which had been on a protein-free diet for three days increased [3]HTdR uptake into DNA at 17 h—an effect which was strongly enhanced by triiodothyronine (see preceding) (Short et al., 1973). If zein, which is deficient in lysine and tryptophan, replaced the casein, the stimulation was not seen. Polysome formation was promoted by administration of amino acids (Clemens & Pain, 1974). In partially hepatectomized rats habituated to a protein-deficient diet with its calorie content maintained, the numbers of mitoses seen in the regenerating livers at 29 and 48 h were about half those counted in the control animals with normal diet ad lib. Even if protein depleted rats were not given casein until 6 h after operation, the usual early increases in hepatic amino acid pool were observed (see following), and the time of the first peak in ornithine decarboxylase activity was unchanged. Poly(A[+])mRNA accumulation in the cytoplasm was, however, less apparent than in normally fed rats (McGowan & Fausto, 1978; McGowan et al., 1979).

Intracellular energy status,

$$\frac{(ATP + \frac{1}{2}ADP)}{(ATP + ADP + AMP)}$$

(Atkinson, 1968), is a further potential limitation on the capacity of cells to respond to growth-promoting signals. It is unlikely, however, that in normal animals where regeneration or compensatory hyperplasia is required, glucose or other energy sources will be limiting. Indeed the facilitating effects of the nonspecific growth-promoting hormones discussed above are likely, in intact animals, to be partly related to their ability to ensure maintenance of energy requirements for cells. Nevertheless, as we shall see later, removal of constraints limiting metabolite entry into cells may be a vital early aspect in the response to the signals.

4.6. LECTINS

Lectins are defined as di- or multivalent carbohydrate-binding proteins which agglutinate cells with complementary saccharide-containing residues in their plasma membranes. The earliest lectins to be intensively studied were isolated from plant sources, but they have now been found in many systems, and receptors for them have located intracellularly as well as on the cell membranes. In *Dictyostelium* the lectin discoidin I may be involved in processes leading to cell-cell adhesion; in vertebrates lectins are being identified which may play some part in muscle development and differentiation. Liver plasma membranes contain a receptor for β-galactosyl residues of asialoglycoproteins which, at the same time, has the properties of a lectin. When circulating asialoglycoproteins are bound onto the receptor, endocytosis occurs (see Barondes, 1981; Jourdian et al., 1981).

If molecules with the properties of lectins are involved in vertebrates in adhesion, it may emerge that they are also involved in tissue reorganization and regeneration. It is clear from model studies on the effects of PHA and conA on lymphocytes (see Section 3.2.2) that lectin binding onto plasma membranes can promote growth, that the sensitivity of cells to such stimulation varies through the cell cycle, and that the number of lectin binding sites also varies (Noonan et al., 1973).

4.7. SIGNALS FOR PROLIFERATION

4.7.1. Conclusions

Regeneration or compensatory hyperplasia occurs postnatally in animals where intercellular communications already operate. The interplay of maintenance hormones—iodothyronines, steroids, insulin, and so on—does not normally produce growth in adult tissues such as muscle, liver, or kidney. Nevertheless, there is good evidence that these hormones facilitate growth when cell or organ depletion has occurred. The changed situation may arise because of alterations in the numbers of receptor sites occupied by normal hormones, so enabling critical thresholds to be exceeded and new responses elicited (e.g., insulin and liver regeneration?). Alternatively, there may be a release of signals specifically associated with the injury, which by themselves produce growth and/or modulate the behavior of their target cells so that growth now follows in the presence of other, normal hormones, for example, EGF.

A few points should be noted:

1. Maintenance hormones are facilitatory—cells in culture or liver cells in endocrinectomized animals will proliferate in the absence of maintenance hormones, although alternative factors may then have to be present.

2. Responses to the signals may be additive. As we shall see (Chapter 5), the signals operate through different transducing mechanisms, which may interact.

3. The characteristics of mitogens are operational, not ubiquitously applicable. Mitoses are produced in specific cell types where a number of criteria may have to be satisfied for cells to move out of G_0, go from G_1 into S, and go from G_2 into M. For different cell types different limitations obtain, and different strategies may be required for their removal.

4.8. SIGNALS PROMOTING DIFFERENTIATION IN POTENTIALLY REGENERATING SYSTEMS

These signals (see Table 4.2) fall into three groups—first, growth factors which cause daughter cells to move out of their growth cycles into G_0 and display phenotypic characteristics associated with their differentiated state, for example, erythropoietin and hemoglobin production by putative red blood cells, and thymosin and surface markers on T cells. The cells affected are already committed to a particular pattern of gene expression. The mechanism by which the factors achieve these effects will be considered in Chapter 5 under Secondary Messengers.

TABLE 4.2. DIFFERENTIATIONAL SIGNALS IN POTENTIALLY REGENERATING SYSTEMS

Agent	System	Response	Reference
Growth Factors			
Erythropoietin	Red cell precursors	Maturation and Hb production	Section 4.3.2
Thymosin	T-cell precursors	T-cell-specific surface markers	Section 4.3.2
Prostaglandins	Macrophages	Surface antigen production inhibited	Snyder et al. 1982
Somatomedins	Myoblasts	Fusion	Ewton & Florini, 1981
Interleukins-1, 2 and antigen	B cells	Antibody formation	Section 4.3.2
Microenvironments			
Marrow	CFU	Proportion granulocytes: erythrocytes	Section 4.3.2
Bone	Osteocytes	Proportion osteoblasts:osteoclasts	Section 4.3.2
Miscellaneous Modifiers			
Ouabain	Erythroleukemic cells	Hb induced	See Marks & Rifkind, 1978, for list of agents active in erythroleukemic cells
DMSO	Erythroleukemic cells	Hb induced	
Hexamethylene bisacetamide	Erythroleukemic cells	Hb induced	
Proteases	Erythroleukemic cells	Hb induced	Scher et al., 1982
Vitamin A	Keratinocytes	Mucus secretion	Fell & Mellanby, 1953
Retinoic acid	Limb regeneration	Restructuring	Maden, 1982
NAD	Chick wing bud	Proportion chondrocytes: osteocytes	Caplan,1981

The second group of factors are involved in determining which, usually of two, alternative lines of differentiation will be followed. Classical and embryological inducers are in this class (e.g., for bone see Bauer & Urist, 1981; Takaoka et al., 1981), and we have already mentioned (Section 3.2.1) microenvironments in marrow and bone which determine the proportions of granulocytes:red blood cells (RBC) or osteoblasts:osteoclasts. The determinants in this class are still undefined; it is usually postulated that the changes are seen only subsequent to DNA replication. Most simply the genome must be exposed in S phase for its expression to be open to modulation.

Third, there are the miscellaneous modifiers—relatively simple molecules (see Table 4.2), which may or may not be important physiologically, that markedly affect phenotype expression.

Two types of differentiated state may be recognized (Section 8.2.3). For neurons, striated muscle, sheets of epithelium, and hemopoietic tissue, commitment to differentiation appears irreversible and the cells move into a G_0 state. For glandular tissues (e.g., adrenal cortical cells, liver) in adult animals, expression of the differentiated traits continues at least through G_1.

A very influential system, in which biochemical processes involved in the attainment of a terminally differentiated state are being studied, is murine erythroleukemic cell lines which can be induced into vigorous hemoglobin production by a wide variety of inducers (Marks & Rifkind, 1978). The cells do not respond to the normal maturation factor for erythroid cells, erythropoietin, but the pattern of changes seen in the induced cells closely parallels normal differentiation. It is not yet possible to identify how the inducing agents work; from their widely various structures (see Table 1 in Marks & Rifkind, 1978) it is implausible that they have a common basis of action. Current biochemical thinking would predict the plasma membrane (Chapter 6), chromatin structure (Chapter 8), and mRNA processing (Chapter 7) to be potential sites of regulation in determined cells. Membrane and chromatin changes are produced by many of the inducers effective in murine erythroleukemic cells; for example, dimethyl sulfoxide (DMSO) (15%) caused an immediate, sixfold increase in the phosphorylation of tyrosine in a 170 kDa protein in liver microsomes (Rubin & Earp, 1983). It is interesting, also, that these agents, which trigger differentiation in the erythroleukemic cells, inhibit proliferation in lymphocytes (Kaplan, 1980).

Induction can be prevented in a number of ways, for example, hexamethylene bisacetamide antagonizes dexamethasone. By alterations in times and levels of exposure to inducer and inhibitor, it has been shown that commitment to hemoglobin production is established sequentially (Marks & Rifkind, 1978). If synchronized cells are used (Marks et al., 1982), hexamethylene bisacetamide must be present at the start of S phase if globin mRNA is to accumulate in the following G_1 period.

Vitamin A and its derivatives are inducing agents whose basis of action, again, is unknown and which have received much attention since the pioneering experiments of Fell and Mellanby (1952, 1953) with embryonic chick wing bud and ectoderm cultured in the presence of excess vitamin A. Initially the vitamin was thought to affect membrane viscosity and permeability (Dingle & Lucy, 1965). More recently other mechanisms are being considered (Ziie & Cullum, 1983). Cytoplasmic binding proteins for retinoids have been described, and the ligand-receptor complex could affect the genome analogously to steroid receptors (Chytil & Ong, 1978). Alternatively retinol phosphate might function as a carrier in glycosyl-transferase reactions, like dolichol phosphate (see De Luca, 1977; Creek et al., 1983), and so affect surface glycosylation profiles and thus intercellular interactions.

Maden (1982) recently showed gross alterations in pattern development in regenerating axolotl limbs if the animals were kept in retinol palmitate for 12 days after operation. The proximo-distal sequence in regenerating limbs was greatly disturbed. Since the presence of the retinoids retarded blastemal development, which recommenced when the animals were placed in pure water, it was concluded that effects were exerted primarily on positional controls and not on cell growth.

The final agent to which attention is drawn (see Table 4.2) is NAD, synthesized in cells from tryptophan and used metabolically and also as mono- and poly-(ADP-ribose) to modulate nuclear membrane and cytoplasmic proteins (Hilz, 1981) (see Section 8.1.3). In chick wing bud cultures, high NAD concentrations have been reported to induce muscle, and low levels, cartilage formation (Caplan, 1981).

Confirmation of these observations has been reported recently from Nishio and colleagues (1983), who used inhibitors of poly(ADP-ribose) synthetase (nicotinamide benzamide) to augment chondrocytic differentiation in the same system.

Understanding how miscellaneous modifiers cause differentiation awaits our understanding of the basis of differentiation. Progress in this area is reviewed in Section 8.2.3.

5

SECONDARY MESSENGERS

Cyclic AMP was discovered and the concept of secondary messengers introduced by Sutherland in the early 1960s. In the 1970s Ca^{2+} also emerged as a secondary messenger. Now cAMP and Ca^{2+} appear as interacting partners (Rasmussen & Waisman, 1981) in the control of a class of protein phosphorylations which is but one of a set of protein phosphorylations affecting different amino acid residues in different sequences, the control being exerted by different extracellular signals. We do not yet know which are the proteins which must be phosphorylated on their seryl, threonyl, or tyrosyl groups for replication and mitosis to occur.

A further complication is that cAMP shows both growth-promoting and inhibiting effects. Its only known means of regulation in eukaryotes is through activation of cAMP-dependent kinase. Many proteins are substrates for this enzyme; at present information is not available for us to decide which proteins are likely to be phosphorylated in different cells. The indications that high levels of cAMP arrest cells in G_1 (see Pastan et al., 1975) suggest that (seryl) protein phosphorylation may control two switches — the first, at relatively low or fluctuating levels, allows cells to progress through their cell cycle, and the second, at maintained, high levels, blocks progress.

Phosphorylation of tyrosine residues has similar dualistic effects. Tyrosine protein phosphorylation by protein kinase (pp60[src]), induced by src viral genes or their equivalent, produces uncontrolled growth in transformed cells (e.g., see Radke et al., 1980; Smart et al., 1981), in contrast to the controlled effects of EGF and PDGF.

Phosphorylation status is unlikely to be the only variable controlling entry of cells into their growth cycle. Ion fluxes, methylation, and acetylation are already implicated. The evidence for these and other possible parameters will be reviewed below.

5.1.   ESTABLISHED SECONDARY MESSENGERS

5.1.1.   Cyclic AMP

LEVELS IN TISSUES FOLLOWING GROWTH STIMULATION
(SEE LLOYD ET AL., 1982)

Regenerating Liver.   Besides the classical, pituitary-hormone-promoted tissues such as the adrenal cortex (Section 4.2.1) and thyroid, variations in cAMP in response to growth were shown very clearly in partially hep-atectomized livers in 1972 by MacManus and his co-workers. Two peaks in cAMP content were detected, the first at about 4 h after operation, and the second at about 12 h. DNA synthesis (^{3}HTdR uptake) reached a peak at 22 h (see Fig. 5.1). Similar observations were reported by Thrower and Ord (1974), who observed an additional, small peak in cAMP content just before the peak of thymidine incorporation (see Fig. 5.1). Analysis with adrenalectomized animals and α- and β-adrenergic blockers indicated that the peaks in cAMP at 4 and 12 h were due to adrenaline release (cf. Brønsted & Christoffersen, 1980). The antagonists had no effect on the levels of cAMP at the G_1/S transition, and this rise was attributed to alterations in insulin/glucagon stimulation after partial hepatectomy (Thrower & Ord, 1975 and Section 4.2.3).

Lymphoid Tissue. Results with thymocytes and mitogen-promoted lymphocytes have been much more controversial than those with regenerating

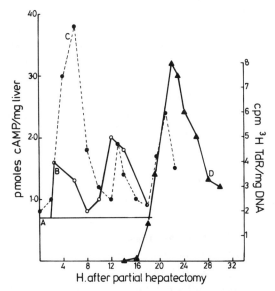

Figure 5.1. Cyclic AMP levels in rat livers after partial hepatectomy. (*a*) cAMP levels, sham-operated animals (MacManus et al.; Thrower & Ord). (*b*) cAMP levels, partially hep-atectomized rats (MacManus et al.). (*c*) cAMP levels, partially hepatectomized rats (Thrower & Ord). (*d*) Relative counts per minute ^{3}HTdR/mg DNA (results coincident for both groups). Reprinted by permission of Academic Press (*Biochemical and Biophysical Research Communications*, Vol. 49, No. 5, p. 1206, 1972) and the Biochemical Society (*Biochemical Journal*, Vol. 144, p. 365, 1974).

liver. Initial demonstrations by MacManus' group (MacManus & Whitfield, 1969; MacManus et al., 1969), showing that dibutyryl cAMP (0.01–1.0 μM) could promote DNA synthesis and mitosis in thymocytes and that concentrations above this level were inhibitory (see Cross & Ord, 1971), were complicated by technical difficulties in measuring cyclic nucleotide levels in mitogen-stimulated lymphocytes and by the high levels of cAMP which are found in cultured cells grown to confluence (see Cell Cultures, following). When these difficulties were overcome, peaks in cAMP were found (Millis et al., 1974) in G_1 and G_2 phases with much lower levels in mitosis. Activity of cAMP phosphodiesterase rose in late G_2 phase.

Cell Cultures.   Those cell cultures, like 3T3 fibroblasts, that are sensitive to density-dependent controls show cessation of growth at confluence; cAMP levels are then high (see Pastan et al., 1975). Usually the cells are arrested in G_1 phase. With CHO cells which had been synchronized by mitotic shaking, reinoculation under growth-promoting conditions confirmed that low levels of cAMP were present in mitosis and that the cyclic nucleotide level reached a peak in G_1, decreasing again as the cells moved into S phase (Sheppard & Prescott, 1972; Russell & Stambrook, 1975; Costa et al., 1976; Green, 1978). What causes the variations in cAMP levels through the cell cycle in cells cultured at low density has not yet been established, but the periodicities may result from down-regulation of receptors, believed to respond to growth factors in the serum in the medium, superimposed on alterations in the constitution of the cell membrane through the cell cycle (Section 6.4).

CONSEQUENCES OF cAMP-DEPENDENT PROTEIN KINASE ACTIVATION

There are two cAMP-dependent kinases, with the same catalytic subunit, C, but different regulatory subunits. Type I regulatory subunit (R_I) is more easily dissociated by low (0.1 M) NaCl concentrations; type II is auto-phosphorylated by the catalytic subunit. When phosphorylated, R_{II} has a reduced affinity for the catalytic subunit (see Geahlen et al., 1982). Both regulatory subunits have additional sites which are phosphorylated in vivo but not by their catalytic subunit. The cAMP-dependent protein kinase preferentially phosphorylates serine residues in sequences Lys/Arg-x-Ser (see Weller, 1979). The conformation and location of the protein also affect its rate of phosphorylation. Frequently proteins which are readily phosphorylated in vitro show insignificant $^{32}P_i$ uptake in vivo.

While resting cells differ in their proportions of type I:type II regulatory subunits, experiments with CHO, liver, and lymphoid cells (Costa et al., 1978; Boynton & Whitfield, 1980; Boynton et al., 1981; Mészáros et al., 1981) indicate that the type II kinase increases at the G_1/S phase transition, probably consequential to new protein synthesis (Costa et al., 1978). Catalytic subunits released from the type II kinase were found in lymphoid and other cell nuclei (Mészáros et al., 1981; Sharma, 1982). The essentiality of these changes and the different roles of type I and type II cAMP-dependent kinases is questionable, however, because CHO mutants with deficiencies in one or another kinase have had the same cell doubling times as wild-type CHO cells (Trevillyan & Byus, 1983).

In liver the most obvious effect of activation of cAMP-dependent protein kinase is glycogenolysis, which is indeed accelerated following partial hepatectomy. In cells in which the glycogen synthetase-phosphorylase system is less evident, it is more difficult to predict which proteins will be phosphorylated by activating the kinase. In 3T3 plasma membranes two proteins (24 and 14 kDa) could be phosphorylated by endogenous cAMP-dependent kinase; the similarity in apparent molecular weights prompted speculation (Scott et al., 1981) that the proteins were identical to those found by Chang and Cuatrecasas (1974) in adipocytes which, if phosphorylated, inhibited insulin-stimulated glucose uptake. It has frequently been speculated that phosphorylation of glucose transport systems is inhibitory; if so, this could be a factor in the growth-inhibitory effects of high cAMP levels.

A number of proteins, for example, tyrosine aminotransferase, were induced in hepatoma cells by cAMP-releasing hormones. Previously it has not been clear if the actions were exerted at the transcriptional or translational level, or both, but with molecular biological techniques this is easier to clarify. In the case of phosphoenol pyruvate carboxykinase (Larner et al., 1982), dibutyryl cAMP increased the concentration of nuclear pre-mRNA for the enzyme five to eightfold within 0.5 h of administration.

Such effects of cAMP on transcription could be brought about in at least two ways—first, by altering the phosphorylation of nuclear proteins and so indirectly facilitating transcription (Murdoch et al., 1982). Alternatively, or additionally, the regulatory subunit(s) could behave analogously to cAMP-binding protein in prokaryotes and promote initiation of RNA polymerase II directly (see Severin & Nesterova, 1982). Which mechanism operates is still uncertain.

cAMP-dependent protein kinase phosphorylates ribosomal proteins in vitro. The difficulties in interpreting such effects and their influence on protein synthesis were discussed earlier (Section 4.2.3).

The cAMP is destroyed in cells by cAMP-dependent phosphodiesterases (EC 3.1.4), of which there are usually two: a low-K_m form in the plasma membrane and a high-K_m soluble enzyme. The low-K_m enzyme may be activated in some cells by an insulin-stimulated phosphorylation (Section 5.2.3); the high-K_m form is activated by calcium-calmodulin. In tissues such as liver, in which many of the relevant parameters are known, attempts have been made to predict the outcome of various extents of stimulation of adenylate cyclase in the presence or absence of insulin and to assess the importance of the different phosphodiesterase isozymes under different conditions (Reynolds, 1982). Intuitively it is evident that, with low levels of activation of adenylate cyclase, the membrane-bound phosphodiesterase is likely to be more important and that little cAMP will escape to diffuse away from the plasma membrane. The range of diffusion of the catalytic subunit of the activated protein kinase, and its sites of action at different levels of cAMP, are not known.

5.1.2. Calcium

CA^{2+} CALMODULIN SYSTEMS

Classical cell biology established that Ca^{2+} was essential in the medium if cells were to proliferate. After the introduction of A23187 as a selective ionophore for Ca^{2+} (see Pressman, 1976), many reports appeared of cell stimulation into growth and proliferation in its presence, reemphasizing the part Ca^{2+} might play in these processes. With the identification of calmodulin as the principle intracellular calcium-binding protein in the range $0.01-1.0$ μM Ca^{2+} (Kretsinger, 1976; Berridge, 1980; Klee et al., 1980; Rasmussen & Waisman, 1981), the messenger role of Ca^{2+} was considerably clarified. In resting cells Ca^{2+} concentrations are in the range $0.01-1.0$ μM. On stimulation the intracellular concentration of free Ca^{2+} rises to $1-10$ μM, exactly in the concentration range in which the ion is selectively and effectively bound by calmodulin. Calmodulin is an acidic polypeptide with 4 Ca^{2+} binding domains/mole (16.9 kDa), which are highly conserved and extensively homologous. Ca^{2+} binding causes marked conformational change in the polypeptide, which is the basis of the regulatory effects of the complex. Since there are four sites for Ca^{2+} in calmodulin, which may or may not be filled, further modulation could arise through different conformations being assumed by the complex at different extents and locations of ligand binding. It has not yet been established whether positive or negative cooperativity is involved in the interactions.

A number of systems can be affected (Cheung, 1980; Klee et al., 1980; Rasmussen & Waisman, 1981). First, cell metabolism is influenced through enzymes such as phosphorylase-b kinase, which are activated by calcium-calmodulin. Other sensitive systems include the high-K_m, cAMP-dependent phosphodiesterase and brain adenylate kinase. Second, there are calcium pumps in the sarcoplasmic reticulum and plasma membrane which are stimulated by calcium-calmodulin. Troponin-C binds two Ca^{2+} in domains which are virtually homologous to those in calmodulin, so initiating contractility in striated muscle. Myosin regulatory light chains may also be affected by a Ca^{2+}-calmodulin-promoted phosphorylation, which may be important in control of certain types of smooth muscle. Microtubular assembly is inhibited by 1.0 μM Ca^{2+}, and alterations in the distribution of calmodulin bound to the mitotic apparatus have been shown immunochemically.

Exocrine secretions are promoted in certain glandular tissues by Ca^{2+}, especially with adrenergic sites (e.g., parotid gland) or acetylcholine muscarinic receptor sites (pancreas). It is postulated, but not yet established (see Rasmussen & Waisman, 1981), that Ca^{2+}-calmodulin may participate in exocytotic processes leading to fusion of the secretory granules with the plasma membrane.

It is evident that some of the effects of Ca^{2+}-calmodulin overlap with and modulate the actions of cAMP. It is not yet clear whether Ca^{2+}-calmodulin activation is invariably accompanied by protein phosphorylation (for discussion see Rasmussen & Waisman, 1981) or whether protein kinase is only one of a set of proteins which can be allosterically promoted by the complex. Calmodulin can be an integral subunit of the system which is then activated by Ca^{2+}, as with phosphorylase-b kinase, or Ca^{2+}-calmodulin complex may bind onto, and so regulate, the system.

These general considerations of the Ca^{2+}-calmodulin system must also include the origins of the increased intracellular Ca^{2+} (see Williamson et al., 1981). In excitable tissues Ca^{2+} enters cells through voltage-sensitive channels consequential on depolarization; in vascular smooth muscle, stimulation of α-adrenergic receptors may allow direct entry.

Ca^{2+} may also be released from sites in the plasma membrane following hormone binding. In pancreas and the salt gland of the albatross, acetylcholine promotes $^{32}P_i$ incorporation into membrane phosphoinositide (Hokin & Hokin, 1955). Michell (1975) has proposed that such hormone binding promotes phosphoinositide hydrolysis, so releasing the Ca^{2+}, which is then free to move into the cytosol. Nishizuka and his colleagues (Minakuchi et al., 1981) have described a protein kinase-C which, in the presence of 1 μM Ca^{2+}, becomes attached to phospholipid, often phosphatidyl-serine, in the plasma membrane and activated. It is postulated that the Ca^{2+}-activated, phospholipid-dependent kinase is coupled to the system causing phosphatidyl-inositol turnover in hormonally stimulated cell membranes. As unsaturated diacylglycerides enhance the activity of protein kinase-C, amplification is possible.

Ca^{2+} can also be released from intracellular binding sites—the sarcoplasmic reticulum in striated muscle—and potentially from mitochondria (see Williamson et al., 1981).

The mechanisms just outlined do not explain growth-promoting effects of Ca^{2+}. Rapid metabolic stimulation occurs (see following) and might have been facilitated by Ca^{2+}. Ca^{2+}-promoted exocytosis may be important in releasing hydrolases which are implicated in the early preblastemic stages of reconstruction. Conceivably such exocytosis might be required in relaying intercellular signals in regeneration. Immunochemical techniques permit calmodulin to be located in many intracellular sites, including within nuclei. Such a location for a Ca^{2+}-calmodulin regulated protein kinase is obviously attractive.

CA^{2+}-CALMODULIN CHANGES IN GROWTH-PROMOTED CELLS (WHITFIELD ET AL., 1980; DURHAM & WALTON, 1982)

Two related hypotheses have been explored: the first that Ca^{2+} is a secondary messenger for various growth factors, that is, intracellular Ca^{2+} concentration rises when growth has been promoted, and the second, that Ca^{2+} itself can promote growth.

Studies with thymic lymphocytes and colony-forming cells from bone marrow, where proliferation was promoted if the extracellular Ca^{2+} concentration rose from 0.02 to 1.0 mM, suggested Ca^{2+} entry might be a factor in lymphoid cell growth (see Whitfield et al., 1980). Technical problems arise because of difficulties in measuring intracellular Ca^{2+} in the μM–nM range (extracellular Ca^{2+} c. 1 mM). Very careful experiments with $^{45}Ca^{2+}$ and conA or A23187-stimulated lymphocytes (Hesketh et al., 1977) failed to detect an increased Ca^{2+} influx. The limits of sensitivity, however, corresponded to 50 μM, and the authors concluded that their result could not exclude the possibility that a small increase in Ca^{2+} was a component in the signal committing the cell to transform. With a nontoxic fluorescent probe a 1.5- to 2.5-fold increase in intracellular Ca^{2+} was detected (resting level 123 nM Ca^{2+}) (Tsien et al., 1982).

The response of 3T3 cells to serum stimulation needs Ca^{2+} in the medium. If the serum is removed, insulin, PDGF, and FGF promote Ca^{2+} inflow and lower the requirement for extracellular Ca^{2+} (Durham & Walton, 1982). Cobbold and Goyns (1983) were able to inject single fibroblasts growing on beads with aequorin and then determine intracellular Ca^{2+} as 0.33 μM. Some 60–70% of cells attached to the beads divided by 38 h after serum addition, and the same proportion (but not, for technical reasons, the same cells) showed a tenfold rise in Ca^{2+}.

In partially hepatectomized rats which had been parathyroidectomized to make them hypocalcemic, DNA synthesis was reduced, although the peak in cAMP was unchanged. From their results Rixon and Whitfield (1976) suggested that calcium deprivation blocked the G_1/S transition.

Immunochemical assays showed a threefold increase in calmodulin in the livers of partially hepatectomized rats in G_1 phase (MacManus et al., 1981); a twofold increase at the G_1/S transition occurred in CHO cells (Chafouleas et al., 1982). In quiescent, serum-deficient cells there was a very rapid fall in calmodulin levels when serum or other mitogens were readmitted, the level again rising at the G_1/S transition. Amounts of calmodulin mRNA paralleled the change in levels of its product.

The basis for the fall in calmodulin mRNA as cells move out of G_0 phase, and its relation to the accompanying rise in intracellular Ca^{2+}, is not known. Nor is it known how the formation of the Ca^{2+}-calmodulin complex enables cells to pass through a restriction point in G_1 phase and initiate DNA synthesis. In the calcium-deprived, partially hepatectomized rats the biosynthesis of deoxyribonucleotides is reduced (see Whitfield et al., 1980), as it is with many other inhibitors (see Ord & Stocken, 1980). What is known about the G_1/S transition in growth-promoted cells is discussed in Section 7.4.

5.1.3. Cyclic GMP

The discovery of cGMP provoked theories that it acted as a secondary messenger analogous to cAMP. Early experiments in growth-promoted cells indicated that levels of the cyclic nucleotide might vary reciprocally to those of cAMP (see Goldberg et al., 1974). Concentrations of cGMP in cells are 0.1–0.01 times those of cAMP in the same cells, and while protein kinases have been described which are preferentially activated by cGMP, the binding constant is not usually sufficiently favorable to preclude more significant activation by the much higher concentrations of cAMP. Measurements of cGMP in, for example, regenerating rat liver showed little variations in levels, at least up to the first wave of DNA synthesis (Fausto & Butcher, 1976).

5.2. FURTHER POSSIBLE SIGNALS OR SECONDARY MESSENGERS

It is not known if all plasma membrane-bound signals require secondary messengers, but other potential messengers have been described, either in adult organisms (methylation—Hirato & Axelrod, 1980) or in early development (NH_3 in *Dictyostelium*—Sussman, 1982) before adult endocrine systems are active. The pattern for limb regeneration is established in the blastema very early, and the strategic similarities between the process in invertebrates and Amphibia (Chapters 1 & 2) suggest that determination of the proximo-distal axis might entail local, sharply attenuating signals from cells in the epidermis and/or amputation plane, effective on the blastema only for a brief period when the blastemal cells are maximally dedifferentiated. During this period signals and secondary messengers may be employed which are no longer used in adult organisms. If the agents used in signaling are lipid soluble or cross the plasma membrane easily by normal transporters, secondary messengers would not be needed if the signaling substances formed part of autocatalytic systems.

5.2.1. Adenosine (Arch & Newsholme, 1978; Fox & Kelley, 1978)

The pharmacological activities of adenosine have been known for some time. It causes vasodilation of coronary vessels; purinergic nerves have been described (Burnstock, 1972), and adenosine may stimulate hormone release. Lipolysis in adipocytes is inhibited by 0.01 μM adenosine, and 0.1 μM blocks platelet aggregation. In many cell cultures 1.0 μM–1.0 mM adenosine is toxic, especially of T lymphocytes.

Extracellular concentrations of adenosine in mammals are in the range $0.1-0.01$ μM; intracellularly they are about $1-10$ μM (Arch & Newsholme, 1978). In many tissues physiological concentrations of adenosine allosterically promote adenylate cyclase, although in adipocytes the enzyme activity is lowered. Circumstances under which adenosine might be released from cells have been described (Arch & Newsholme, 1978) (see Fig. 5.2). Elevated concentrations of adenosine in extracellular fluid have been reported after serious tissue injuries, such as those following total body irradiation. Locally increased levels might be expected during the preblastemic period because of tissue destruction.

Similar considerations in respect of $2'$-deoxyadenosine can also be advanced (Fox & Kelley, 1978).

5.2.2. Ion Fluxes

Transient depolarization of the plasma membrane is believed to be a very early consequence of growth stimulation by many agents. It can be monitored directly with microelectrodes (see Rothenberg et al., 1982) or indirectly with potential-sensitive fluorescent dyes (e.g., see Felber & Brand, 1983) or by measuring the distribution of lipophilic ions. Depolarization of the plasma membrane has commonly (see Moolenaar et al., 1981; Kiefer & Schulze, 1982; Rothenberg et al., 1982) but not invariably (Deutsch & Price, 1982) been observed early after stimulation. Some differences in membrane response may arise because cells are compared in different initial

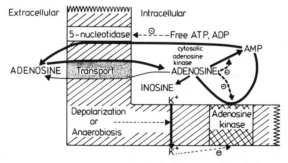

Figure 5.2.   Potential mechanisms for the control of adenosine release from cells (Arch & Newsholme). Reprinted by permission of the Biochemical Society (*Essays in Biochemistry*, Vol. 14, p. 97, 1978).

states. ConA causes rapid depolarization in pig lymphocytes, but in resting mouse thymocytes, hyperpolarization is produced, Ca^{2+}-dependent K^+ channels being stimulated (Felber & Brand, 1983). Depolarization is usually accompanied by Na^+ entry, which may be complex in origins, having electrogenic and electroneutral components (Rothenberg et al., 1982; Felber & Brand, 1983). One of the electroneutral mechanisms is for Na^+ entry to be balanced by H^+ extrusion, resulting in a rise in intracellular pH, as has been detected in 3T3 cells (Schuldiner & Rozengurt, 1982) and in BSC-1 epithelial cells following addition of serum or EGF (Rothenberg et al., 1982).

The inhibition of cell proliferation by ouabain (blocks plasma membrane Na^+-K^+ ATPase—"Na^+ pump"), monesin (a Na^+ ionophore), and amiloride (inhibits Na^+-specific transport and protein synthesis, Lubin et al., 1982) in 3T3 cells (Rozengurt & Mendoza, 1980; Lubin et al., 1982) and rat primary hepatocyte cultures (Leffert & Koch, 1980) confirms that monovalent ion fluxes are involved in the growth response.

K^+ influx may also be observed. Studies with $^{42}K^+$ or with $^{86}Rb^+$, which mimics the behavior of K^+, showed entry in 3T3 cells to be stimulated by serum, prostaglandins, EGF, and insulin (Rozengurt & Mendoza, 1980); PHA promoted K^+ influx in lymphocytes (Kaplan & Owens, 1980). It is possible that increased K^+ entry is consequential on increased intracellular Ca^{2+}, which opens Ca^{2+}-dependent K^+ channels (see Felber & Brand, 1983).

In other cells early G_1 is associated with net efflux of K^+, followed by net influx at G_1/S (de Laart & van der Saag, 1982). Conflicts arise because of differences in the relative importance and interactions of the various processes involved in monovalent cation uptake in different cell types.

There are difficulties, too, in interpreting the inhibitor studies. Constancy in intracellular Na^+ and K^+ concentrations is part of normal homeostasis and, as such, is essential to proper cellular functioning. Na^+ is cotransported across the plasma membrane with various solutes which enter cells by mediated transport, for example, many amino acids. Increased uptake of solutes is frequently observed as cells move out of G_0 and is thought to be at least facilitatory for growth (Section 7.1.2). Interaction with H^+ pumps can occur, and Na^+ and K^+ ratios may affect the selection of messages for translation (see Burgoyne, 1978), that is, many stages in cell metabolism may interact with Na^+ and K^+ transport.

Further, the clear demonstrations that monovalent cation fluxes are stimulated by many of the growth-promoting factors may be secondary to other actions of these hormones. For example, insulin may stimulate the Na^+ and K^+ pumps (Gavryck et al., 1975; Moore, 1983).

The interdependence of metabolite transport, Na^+ and K^+ fluxes, and alterations in hydrophobicity and charge distribution consequent on ligand binding by the plasma membrane makes it difficult to establish causality in this area. Operationally, binding growth-promoting peptides to plasma membranes induces a set of changes in the cell (e.g., Na^+ entry, depolarization, increased amino acid transport, increased intracellular Ca^{2+}, increased protein kinase activity) which are coordinated, interacting, usually include at least one amplifying stage, and may be self-regulatory (Hershko et al., 1971). Inhibition of any part of this coordinated response can prevent growth.

5.2.3. Methylation

Methylation of phosphatidyl ethanolamine may increase membrane fluidity. Plasma membranes from a number of cells have bound methyl transferases, which may be activated by hormone binding to β-adrenergic receptors. Increased lipid methylation increases the sensitivity of adenylate cyclase to β-adrenergic agonists. Increases in lipid methylation have also been reported after other ligand binding, for example, conA, but are not invariable (see Hirato & Axelrod, 1980).

In addition to phosphatides other macromolecules can be methylated. In many, but not all, species, methylated DNA is associated with transcriptionally inactive chromatin (see Razin & Friedman, 1981 and Section 8.1.3); the pattern of methylation is probably laid down very early in embryonic development (see Jähner et al., 1982) and perpetuated (Wigler, 1981). The r and t RNAs are also methylated, as are proteins (Paik & Kim, 1979). With histones, although variations in the extent of methylation were originally postulated, it is now thought that most methylation occurs shortly after the proteins are synthesized and is then invariant.

As far as is known at present, activities and amounts of methylases do not alter in growth-promoting conditions. S-adenosylmethionine is an important and very active metabolite, but although methionine is one of the less common amino acids, there is little evidence of its limited availability to extrahepatic tissues. Nevertheless the possibility remains open that methylation may serve as a significant intracellular signal, alterations in its extent being elicited by conformational changes in its substrate at some stage in the cell cycle.

5.2.4. Other Protein Phosphorylations

Phosphorylations dependent on cAMP (Section 5.1.1) and Ca^{2+}-calmodulin (Section 5.1.2) have already been described, as have tyrosine phosphorylations (Section 4.2.2). Several other protein kinases are now becoming characterized (Cohen et al., 1982), notably casein kinases I and II (Dahmus, 1981; Hathaway & Traugh, 1982; Pierre & Loeb, 1982), the very similar, possibly identical, enzymes from nuclei (nuclear kinases I and II) (Thornburg & Lindell, 1977; Baydoun et al., 1981; Yutani et al., 1982), and another nuclear kinase which extensively phosphorylates histone H1 (see Matthews, 1980). Casein kinase I preferentially phosphorylates serine (Ser) in the sequence Glu-x-Ser, where x = leucine (Leu), glutamate (Glu), or asparagine (Asn), and casein kinase II preferentially phosphorylates both serine and threonine with acidic residues on the carboxy terminal side of the hydroxy amino acids (Meggio et al., 1981; Hathaway & Traugh, 1982). Their natural substrates are still uncertain but include eukaryotic initiation factors in vitro (Hathaway et al., 1979). Casein kinase II is activated by polyamines and inhibited by heparin (Meggio et al., 1982). Ornithine decarboxylase activity is frequently rate limiting for the production of polyamines. The enzyme is induced when growth is promoted (see Russell & Haddox, 1979 and Section 8.2.3) but may also be activated by release from an inhibitory protein antienzyme (Heller et al., 1976; Viceps-Madora et al., 1982) which may be a nuclear, polyamine-dependent protein kinase (Atmar & Kuehn, 1981), presumably nuclear kinase II. Polyamines can potentially exert marked effects on cellular organization and function (Section 7.2.3). The possibility is now emerging that the bases are part of a feedback regulated protein phosphorylation system, catalyzed by casein kinase II (Atmar & Kuehn, 1981).

Functions of casein kinase I are still uncertain, but its inhibition can be correlated with a block in growth and steroid-promoted enzyme induction in rat liver (Ord & Stocken, 1981a; Cummings et al., 1982).

Increasing numbers of peptides are being isolated which inhibit protein kinases (see Guasch et al., 1982). Their physiological significance is not always clear. It is not yet known if all protein kinases are regulated.

It is evident from this discussion that, although most protein phosphorylations are not highly specific, subsets of proteins are now emerging which are potentially affected by one (or more) kinase. The phosphorylated protein substrates can be distinguished by sodium dodecylsulfate polyacrylamide gel electrophoresis. In a recent study with Chinese hamster lung fibroblasts, Chambard and colleagues (1983) identified three proteins whose phosphorylation appeared to be essential as cells moved out of G_0—ribosomal protein S6, a nuclear protein of M_r c. 62 kDa, and a cytoplasmic 27 kDa protein. Serum, thrombin, FGF, and PDGF all promoted DNA synthesis in these cells and caused the three proteins to be phosphorylated. Insulin and IGF did not elicit phosphorylation of the 27 kDa substrate.

It is often very difficult to establish with certainty which phosphorylation systems are involved and whether amounts or activities of the various enzymes are changing. Choice of substrates also influences the interpretation of the results unless kinase-specific peptides are used. Nevertheless, in studies of phosphorylation in regenerating newts limbs, variations in phosphorylation were detected between the newly established blastema and the stage when digits were reformed. Increases in K_m of cAMP phosphodiesterases were unequivocal (Carroll & Sicard, 1980; Laz & Sicard, 1982).

Of course, we need to know not only which proteins are phosphorylated, but what function they exert. If multiple sites of phosphorylation are present within the same protein, catalyzed by phosphorylating enzymes of different specificity, subtle interactions are possible, especially if the substrates are themselves enzymes (e.g., for glycogen metabolism see Cohen, 1980).

5.2.5. Other Protein Modifications

Methylation of macromolecules and phosphorylation of protein are but two of an increasing list of modifications to intracellular proteins which have now been reported (see Table 5.1). Many of these are important in vivo in specialized cells, for example, carboxylation of preprothrombin in liver and hydroxylation of proline in fibroblasts. They may be regulated, for example, histone acetylation, which is variable and correlated with tran-

TABLE 5.1. COMMONER POSTTRANSLATIONAL MODIFICATIONS TO PROTEINS

Modification	Acceptor Amino Acids	Donating Agent	Protein Substrate/ Comment	Reference
Acetylation	Lys	Acetyl CoA	Core histones, etc.	see text
Amidation	Asp, Glu	Gln, (NH_3?)	Widespread	see Stryer, 1981; Rimmington & Russell, 1982
Carboxlyation	Glu	CO_2/vit K	Preprothrombin	Olsen & Suttie, 1977
Glycosylation[a]	Hydroxy amino acids	UDP-sugars	Extrinsic proteins in plasma membranes:	Stryer, 1981; Hubbard & Ivatt, 1981
	Asn	Dolichol-PO_4^--sugars	secreted proteins	
Hypusin[a]	Lys	Spermidine	Amount increased in growth-promoted lymphocytes	Cooper et al., 1982
Hydroxylation	Lys, Pro	O_2	Preprocollagen	Wold, 1981
Methylation[a]	Arg, His, Lys	S-adenosyl-methionine	Many	see text
	Glu/Asp	S-adenosyl-methionine	Carboxymethyl transferase stimulated in chemotaxis	O'Dea et al., 1978
O-phosphor-ylation[a]	Ser, Thr, Tyr	ATP	Many	see text
N-phosphor-ylation[a]	His, Lys	ATP	Very unstable	Chen et al., 1977
Adenylation	Tyr	ATP	Glutamine synthetase in E. coli	Chock et al., 1980
Uridylation	Tyr	UTP	Glutamine synthetase	Chock et al., 1980

ADP-ribosylation[a] (mono & poly)	Arg, Glu, Ser-PO$_4$	NAD	in *E. coli* Many proteins	Purnell et al., 1980; Hilz, 1981; & text
Proteolysis			Irreversible, widespread	
Redox changes[a]	Cys		Spindle apparatus	see Mazia, 1961
Mixed disulfides	Cys	GSH	Albumin	Freedman, 1979
Sulfation	Tyr	Cys/Phospho-adenosine phosphosulphate	Detected in cell cultures	Huttner, 1982
Ubiquitinization[a]	Lys	Ubiquitin	Conjugated to histone H2A	Goldknopf & Busch, 1977

Source: Wold, 1981.
[a] growth associated.

scriptionally active chromatin, possibly, though not certainly, through controlled deacetylation (Sections 8.1.3 & 8.3.1). In some cases (adenylation
of glutamine synthetase), regulation is well established but restricted to a
limited number of cell types, for example, *E. coli*. Many of the examples
are not obviously linked to extracellular signals. Often the modifying agents
are not rate limiting, or they are only restricted in amounts in profoundly
"switched off" cells such as unstimulated small lymphocytes. As we have
mentioned, the variable factors may be neither the modifying agent nor the
activity nor the amount of the modulating or demodulating enzymes, but
the conformation of the protein substrate (see discussion in Fonagy et al.,
1977).

Some very simple substances, such as NH_3, CO_2/HCO_3^-, and H^+, are
potential modifiers and signals. They could be important in embryogenesis
or in setting up the developmental program in a blastema where adult intercellular communication systems through nerves or the circulation have
not yet been reestablished. Modifications are also good candidates for intracellular signals for phase changes in the cell cycle.

Extreme caution is needed in this field. Isotopic methods make it relatively
easy to demonstrate modifications in isolated systems or in cell cultures in
vitro. It is often more difficult to assess modifications in vivo because the
pool size of the modifying agent may be relatively large, leading to marked
isotope dilution (e.g., phosphate, acetate). The extent of modification on
a macromolecule can also be small, though significant (e.g., 1 amino acid/
100 residues). If the substrate is already modified and the labelling period
is short compared with the half-life of the modification, little isotope incorporation will be measurable. Even if a modification is detected and can
be correlated with physiological changes, causation or consequence is extremely difficult to sort out. It has already been noted that phosphorylation
systems, for example, interact. Neither the discovery of specific inhibitors
for a modification nor the development of mutants necessarily, therefore,
produce "leak-proof" lesions. Further, we know that at least some of the
interesting mechanisms, for example, DNA methylation and cAMP-dependent
kinases, are not universally employed in eukaryotic cells.

A further serious problem is that, even where correlations are found
between the extent of a modification and a particular intracellular state, for
example, phosphorylation of histone H1 at G_2/M (Section 8.4.1), it is often
difficult with structural proteins, or with those whose function is still unknown
(e.g., many nonhistone chromosomal proteins), to establish that the mod-

ification has any biological significance. In spite of these qualifications teleological thinking is persuasive that macromolecular modifications matter.

5.3. TRANSCRIPTIONAL CONSEQUENCES OF SECONDARY MESSENGERS

The altered patterns of gene expression required in growth and tissue restructuring are elicited by controls with relatively low specificity, such as the different classes of phosphorylation. Highly specific controls analogous to prokaryotic repressors may be used in blastulation when the differentiating program is determined and the potential for gene expression laid down. Thereafter alterations in transcription would be brought about:

1. By direct interaction of secondary messengers or their products with regulatory regions on DNA, as for steroid hormone receptor binding to promoter sites (Section 8.2.1).

2. Indirectly by modification of proteins adjacent to the regulatory regions, the potential availability of these proteins as substrates having been determined by the organization of the differentiated genome during embryogenesis (Section 8.2.3).

3. By alterations in the intracellular environment, promoting transcriptional switching or alternative posttranscriptional or translational processing (Section 8.2.3).

6

THE RESPONSES— THE MEMBRANE

There are three facets to the involvement of the cell (plasma) membrane when growth is promoted: its reception of the signal, the changes the signals produce in the properties of the membrane to enable the cell to move out of G_0 into the growth cycle, and the information the outside of the cell membrane imparts to other cells for tissue assembly and organ restructuring.

6.1. SIGNAL RECEPTION

While we are primarily concerned with the way in which the membrane responds to intercellular signals to amplify the effects of their binding and to transduce these into secondary intracellular effects (Chapters 4 & 5), two other aspects of membrane properties must be considered. Hormones frequently affect the transport of simple metabolites. We will also examine clustering of the ligand-receptor complexes, as well as the highly specialized capping responses of lymphocytes and other cells to antigens and mitogens (Weatherbee, 1981), because of the clues which are emerging linking ligand binding to hormone receptors with changes in the cell cytoskeleton.

6.1.1. Basic Membrane Structure and Ligand-Receptor Interactions (Schulster & Levitski, 1980; Houslay & Stanley, 1982)

In simple terms we know the membrane is composed of a lipid bilayer containing integral proteins, some of which may span the membrane and others which only project into the lipid bilayer with part of the protein on the extra- or intracellular side. Additionally, predominantly peripheral hydrophilic proteins bind to the polar surfaces of other membrane proteins and/or lipids. On the extracellular face of the cell, both proteins and lipids may be glycosylated; all hormone receptors so far characterized are glycosylated proteins. With an average mass ratio of lipid to protein of 1:1, there are about 50 molecules of lipid to one of protein.

There is considerable mobility of molecules within the membrane—the phospholipids have a fast axial rotation ($D_{axial} = 10^{-8}$–10^{-9} s) and lateral diffusion (10^{-8} cm^{2}s^{-1}), as well as a slow exchange (the so-called flip-flop taking up to a day) between the two halves of the bilayer. Integral proteins also possess lateral and rotational motion, but while some are extremely mobile, others have little or no lateral mobility, probably due to association with each other or with the cytoskeleton (Weatherbee, 1981). Once integrated the mobility of the protein is affected by the type of lipid with which it is in immediate contact—the lipid annulus.

Membrane changes occur during the cell cycle; microviscosity is maximal during mitosis and minimal in the S phase. Changes are observed when lymphocytes are triggered by mitogens, due to the clustering of the more rigid sphingolipids and gangliosides within the membrane. The activities of the adenylate cyclase, Na^+-K^+ ATPase, and the Ca^{2+} gates are also promoted.

Increased fluidity (see Houslay & Stanley, 1982) is a characteristic of the plasma membranes of tumors and regenerating liver, and an increased fluidity following stimulation of quiescent cells has been correlated with a decreased cholesterol:phospholipid ratio in plasma membrane from regenerating or malignant hepatocytes.

Various strategies are used to couple signal reception with transduction. The simplest, exemplified by the receptor for EGF, is that a permanent association exists between the receptor and its transducing protein on the cytoplasmic side of the membrane—in this case the tyrosine kinase subunit. The multicomponent, direct coupling model for the hormones interacting with adenylate cyclase has already been considered (see Section 4.2.1). Alternatively, in the fluid model (see Fig. 4.1) the receptor or the ligand-receptor complex is mobile, and activation only takes place if the ligand-receptor complex binds to the transducer long enough for activation to occur. With this model the ligand-receptor complex could activate several units before dissociation, so providing an amplification mechanism for the stimulatory ligand.

In some cases it seems that a one-to-one combination of ligand and receptor is inadequate to promote activation; mitogens, for example, are bivalent. Insulin binds with high affinity to receptor glycoproteins on the cell surface. Studies with antiinsulin receptor antibodies (Section 4.2.3) have established that these will mimic the actions of insulin if bivalent antibodies are used to interact with the receptors, suggesting multivalent insulin-receptor interactions are needed to elicit a response. The combinations then form clusters which are internalized. EGF clusters and internalizes its receptors, and this seems to be critical for their biological activity, since derivatives of EGF which bind with a high affinity but do not cluster have no activity. Activity can be restored by cross-linking the derivatives with EGF antibody, which is a divalent reagent. It is not clear if internalization is required for the response, but it is evident that endocytosis is a means by which the number of the receptors, and hence the extent of the stimulus, could be regulated.

Information on how the signals are received and transmitted by the plasma membrane can be obtained from the properties of the isolated membrane and from reconstitution experiments in which functional proteins are inserted into defined lipid vesicles or bilayer membranes. For example, vesicles or artificial membranes have been constructed which can transport amino acids or respond to acetylcholine. The conformational changes in the receptors still elude us. The isolation of sufficient receptor protein in its native form, followed by its insertion in the membrane bilayer with proper orientation, is the main technical barrier to further advance. So far it has not proved possible to reconstruct hormone-sensitive receptors coupled to their transducers.

6.1.2. Clustering and Capping

The questions of what are the physical and physicochemical changes in the plasma membrane when the quiescent cell is stimulated by mitogens have not yet been completely answered. Observations with lymphocytes showed that antigen receptors, which are diffusely scattered over the membrane surface, clustered into patches when the ligand was bound; these patches then aggregated further and formed a cap over one region of the cell. This cap was subsequently endocytosed and the contents of the vesicle released inside the cell (Schreiner & Unanue, 1976; Weatherbee, 1981). In the study of the capping in L cells (Otteskog et al., 1981), it was noted that the cap was localized opposite the nucleus, which was displaced to the periphery of the cell, suggesting that the nucleus might have a role to play in the process.

Capping requires cross-linking of surface ligands. Monovalent F_{ab} fragments of antibodies do not produce capping, but with cross-linking with an anti-(F_{ab}) antibody, capping ensues. Capping has been shown to occur in many different cells as well as lymphocytes—normal and transformed fibroblasts, CHO cells, macrophages, rabbit ovarian granulosa cells, and lymphoblastoma cells (Weatherbee, 1981). Interestingly, periodate induced transformation of human peripheral blood lymphocytes (Banchereau et al., 1981). Presumably the vicinal hydroxy groups of fucosyl, sialyl, and galactosyl residues were oxidized to give free aldehyde groups on receptor molecules, which could join by aldol condensation to form aggregates.

When lymphocytes are challenged with antibody or conA, calmodulin, which in resting cells is diffusely distributed throughout the cell, concentrates in the region just below the cap, provided Ca^{2+} is present in the external medium (Salisbury et al., 1981). Additionally it has been shown by immunofluorescence microscopy and the simultaneous use of fluorochromes that at early times after ligand binding there was an accumulation of actin, myosin, and α-actinin under the patches (Weatherbee, 1981). These data suggest that capping results from an active process depending on the interaction of the contractile proteins with the ligand-receptor complexes in the membrane. Other evidence for this derives from the isolation of complexes from normal and capped lymphocytes. In the capped cells there was a much greater proportion of the surface immunoglobulin associated with actin (Flanagan & Le Koch, 1978).

Explanations of how capping takes place have been evolving as our knowledge grows of the components involved (Bretscher, 1976; Berlin & Oliver, 1978; Braun et al., 1978; Hewitt, 1979; Koch, 1980). Most authorities accept that the concentration of ligand-receptor complexes by clustering and capping is associated with their interactions with underlying actin microfilaments which are part of the cytoskeleton (see Heath, 1983). It has also been shown (Füchtbauer et al., 1983) in cultured cells that the capping proteins disrupt the microfilament bundles, causing their disintegration from the distal end towards the center of the cell. A multiplicity of proteins bind to actin; the binding is highly sensitive to Ca^{2+} concentration (Weeds, 1982). Since the Ca^{2+} gates in the plasma membrane are sensitive to changes in fluidity produced by the mitogenic stimulation, changes in Ca^{2+} could alter the association of actin, myosin, and other proteins and hence facilitate communication between the receptors on the membrane and the structural elements of the cell. How far these events could influence the nucleus, which is presumably the prime target for any stimulant, is still undetermined. Microfilament bundles made up of contractile proteins have been identified in a variety of cell types, and Heath and Dunn (1978) have observed that the bundles terminate in a loose matrix of microfilaments surrounding the nucleus.

These observations of clustering and capping may also be related to the movement of cells, which is an essential feature of blastema formation and subsequent restructuring of the tissues. Phagocytosis is associated with a concentration of microfilaments to form pseudopods which engulf particles attached to the cell surface. The changes in microfilament distribution are analogous to those following capping but are localized to the region of the membrane where the particle is being phagocytosed.

6.1.3. Changes in Membrane Transport following Growth Promotion

GLUCOSE

Mediated transport across the plasma membrane is frequently facilitated by growth-promoting factors. Effects are often immediate (within minutes) and independent of RNA synthesis (see Section 7.1.2) but may be followed after some hours by additional cycloheximide-sensitive enhancement. Insulin increases glucose uptake in many cell types by recruiting new transporter proteins to the plasma membrane (Lienhard, 1983). These proteins are contained in vesicles inside the cell, but the mechanisms of the translocation of the carrier molecules into the membrane (see Houslay & Stanley, 1982) are, at present, speculative.

AMINO ACIDS AND PYRIMIDINE RIBOSIDES

It is not known whether transporters for other solutes are recruited similarly to those for glucose; increased amino acid and pyrimidine riboside transport is certainly observed in many tissues which are promoted into growth in vivo (see Section 7.1.2 and Shotwell et al., 1983). Two processes are seen—an immediate increase in transport independent of protein synthesis, and a slowly developing response (Bonghetti et al., 1978) which is actinomycin sensitive and probably related to membrane protein synthesis as cells move through G_1 phase (Section 6.3). In liver, hemodynamic factors may play some part in the response (Section 7.1.2), and for amino acids (e.g., phenylalanine uptake in PHA-promoted lymphocytes, Kay, 1976) Na^+-dependent transport (see Table 6.1) could be promoted following increased Na^+ entry (Section 5.2.2).

Primary cultures of rat liver hepatocytes grown as monolayers showed an increased Na^+-dependent transport of aminoisobutyrate when challenged with insulin. There was a brief lag period of about six hours after addition of the insulin before the changes could be detected. Inhibitor studies indicated that RNA and protein synthesis were required for the effect. If the cells were grown in suspension culture, stimulation by insulin of amino acid uptake was not seen, illustrating the importance of cell shape/surface area

**TABLE 6.1. COMMON AMINO ACID TRANSPORT SYSTEMS IN
ANIMAL CELLS (See Guidotti et al., 1978)**

System	Amino Acids	Na$^+$ Dependency	Distribution and Comment
A	α-Aminoisobutyric acid, short, polar, or linear side chains, tolerates N-CH$_3$ groups	+	Mesenchymal and epithelial cells, not RBC and reticulocytes
Glycine	Glycine	+	RBC and reticulocytes
Imino-glycine series	Gly, Pro, hydroxypro	+	Renal tubule
ASC	Ala, Ser, Cys, not N-methylated amino acids	+	Ubiquitous, including RBC and reticulocytes
L	Branched or ring amino acids	No	Strong exchange properties
Ly$^+$	Lys, Arg	No	Strong exchange properties

in the response (Kleitzen et al., 1976). Further studies by Handlogten and Kilberg (1982) showed that Na$^+$-dependent transport of neutral amino acids by system A (see Table 6.1) was stimulated by insulin, glucagon, or dexamethasone when tested on hepatocytes in monolayer cultures isolated from newborn and adult rats. There was no such response if hepatocytes from fetal (14–18 day) animals were used (see Section 4.2.3).

6.2. GAP JUNCTIONS

An alternative device, by which cells can receive signals directly from their neighbors rather than via the extracellular environment, is the gap junctions (connexons) (Robertson, 1963; Loewenstein, 1979; Hertzberg et al., 1981). These gap junctions are seen in the electron microscope as a hexagonal lattice of cylindrical particles with a stain filled core. Each assembly is aligned with a similar unit in an adjacent cell to form a channel of 18–20 Å diameter which allows the intercellular diffusion of small molecules with molecular weights of less than about one kDa. Unwin and Zampighi (1980), from an electron microscopic study of isolated rat hepatocyte junctions, have suggested a simple molecular mechanism by which the permeability could be controlled in vivo. The model consists of six rod-shaped subunits which span the membrane and, by appropriate rotation, can open and shut the channel. The chemical composition of the junction is not completely agreed. Henderson and colleagues (1979) have isolated two proteins ($M_r = 26$ and 21 kDa) with sequence homologies; the smaller protein was possibly derived from the larger by proteolysis. The proteins form multimers, and the authors believe the junction consists of only one (26 kDa) protein, present as a hexamer. Cholesterol is tightly bound to the protein, whose hydrophilic residues face the inside of the channel. The major polypeptide has a half-life of about 5 h in mouse liver, in contrast to an average of 23–25 h for other proteins in the plasma membrane. The availability of the protein might be a means by which intercellular communication by gap junctions is controlled (Fallon & Goodenough, 1981).

In adults, junctions have been observed in all tissues except muscle and nerve, and in the embryo, channels are present even in these. Junctions can be made between homologous and heterologous species. Hamster fibroblasts from kidney join with fibroblasts from the mouse, but homologous connections form most readily. The further apart the species are phylogenetically, the less likelihood there is for junction formation.

These gap junctions provide an excellent means by which the transfer of small molecules, such as nucleotides, metabolites, and small hormones, can be effected. Material can pass in both directions, and since the channels do not permit loss to the extracellular space, growth-controlling information can be disseminated through a cell population with minimum dilution. A potential for molecular sieving makes gap junctions especially interesting as a control device. It is possible to reduce the annulus gradually in vitro and so discriminate between the size of the molecules transferred. The annulus responds to changes in Ca^{2+} concentration, and if the normal physiological concentration of 0.1 μM is raised to 10 μM, the transit of all but the small inorganic ions is blocked (Rose et al., 1977).

Loewenstein and co-workers have investigated the transfer of small molecules in cultured mammalian cells and the influence thereon of cAMP (Flagg-Newton et al., 1981). The peptides Glu-Glu-Gln (Gln for glutamine), Glu-Gln-Glu, or Leu-Leu-Leu-Glu-Glu were used as probes and the cells exposed to cAMP, dibutyryl cAMP, or caffeine. There was a rise in the number of gap junctions and an increased transfer at 4 h (Flagg-Newton & Loewenstein, 1981). An increase in serum concentration caused more junctions to form, whereas an increase in cell density produced a fall. C11D mouse cancer cells growing in culture do not make permeable junctions, but if cAMP is added they do (Azarnia et al., 1981).

Constraints in establishing gap junctions are indicated from the experiments of Warner and Lawrence (1982), who studied cell-to-cell communication between epidermal cells of fifth instar larvae of the milkweed bug, *Oncopeltus fasciatus*, and of blowfly maggots, *Calliphora erythrocephala*. There was complete communication of ions, lucifer yellow (M_r c.0.45 kDa), and lead EDTA (0.374 kDa) between cells in the same segment but not of lucifer yellow between segments, suggesting that gap junctions at the segmental border have different properties than those in the same segment. This is particularly interesting as it is already well known that cells are committed very early to a particular segment in development. Cells in one compartment never give rise to cells in an adjacent compartment (Crick & Lawrence, 1975). Since there is no specificity in the way the channels allow the passage of molecules below a given size, if the junctions are involved in pattern formation in tissue restructuring, it could only be by their transmission of material from a localized source. The supply of information generated from such a source could be either intermittent or continuous, giving rise to an oscillating or uniformly declining gradient (see Loewenstein, 1979).

6.2.1. Gap Junction Changes in Regeneration

During would healing following removal of 4 cm of skin in rats, there is an increase in the proportion of the cell surface occupied by gap junctions between myofibroblasts (Gabbiani et al., 1978). On the other hand, Meyer et al. (1981) found the area of the hepatocyte membrane occupied by gap junctions to be reduced a hundredfold at 29–35 h post partial hepatectomy. In weaning rats Yee and Revel (1978) also found that the number of gap junctions was unchanged between zero and 20 h following partial hepatectomy but was markedly reduced at 28 h at the peak of mitosis. This was confirmed by Traub and colleagues (1983), who measured the amount of the 26 kDa component immunochemically and found it to decline to 15% of normal levels. By 48 h the surface area occupied by the junctions, approximately 3%, was back to normal. In adult livers each cell had channels to six others but in the regenerating liver to only one. Possibly, in the early stages of regeneration, direct interhepatocyte communication is relatively unimportant, or perhaps it takes a significant time for gap junctions to be made. Whatever the reason, the decline in gap junctions during the early stages of liver regeneration suggests that the radially distributed, synchronous pattern of mitoses seen in parenchymal cells across the lobules from the periportal vessels (Section 3.3.1) has not been significantly ordered by metabolite communication through the gap junctions.

6.3. SIGNALS AND THEIR RECEPTION: SUMMARY

The discussions of intercellular signals, their reception by the cell membrane, and their transduction into intracellular effects (Chapters 4 & 5) can be summarized in postulates which can be offered about intercellular signals for regeneration:

1. More than one signal is likely to be required, in addition to normal endocrine involvements in homeostasis (Section 4.7.1). Tissue restructuring affects several cell types, which must be specifically addressed, stimulated to proliferate, and appropriately organized spatially.

2. Because of these specificity requirements, the signals will be peptide in nature, binding onto high-affinity receptors in the cell membrane (Section 4.2).

3. Higher-molecular-weight signals are likely to be oligomeric and show multivalent binding to more than one receptor. Hormone receptor clustering and endocytosis frequently follow, allowing down-(auto)regulation of the cellular response to the signal (Section 6.1.2).

4. The response is pleiotropic (See Table 9.1). Binding of the signal to the membrane receptor leads directly or indirectly to an increased Na^+ entry, H^+ extrusion, and a rise in intracellular pH (Section 5.2.2), a rise in intracellular Ca^{2+} (Section 5.1.2), and an increased entry into the cell of solutes, notably glucose, some amino acids, and pyrimidine ribosides, which are normally translocated by facilitated diffusion (Section 7.1.2).

5. Intracellular effects are likely to include a metabolic boost. ATP concentrations may rise or the metabolic flux might be enhanced. Protein synthesis will be increased and the processing of pre-mRNA altered so that more mRNA reaches the ribosome (see Chapter 7).

6. These intracellular changes almost certainly involve protein phosphorylation as a regulatory device (Sections 5.1.1, 5.1.2, & 5.2.4).

7. Tyrosine phosphorylation in an as yet unidentified protein is a critical requirement for mitogenesis (Section 4.2.2).

6.4. CHEMICAL AND MORPHOLOGICAL CHANGES THROUGH THE CELL CYCLE

There is an extensive literature describing the morphological changes which take place during the cell cycle. Cells in monolayers round up in mitosis, while cells in suspension remain spherical throughout the cycle. Electron micrograph studies by Abraham and his colleagues (1973), Porter and colleagues (1973), and others (see Lloyd et al., 1982) have shown the presence of ruffles, microvilli, and blebs in G_1. In S phase cells become smooth, and in G_2 the number of microvilli increases (Porter et al., 1973; Knutton, 1976), but in the case of kangaroo rat cells (Sanger & Sanger, 1980), the microvilli are lost as mitosis begins, only to reappear at the beginning of telophase. Sanger and Sanger have suggested that the microvilli serve to arrange the actin filaments, to generate a contractile force for the rounding up of the cells in mitosis.

Components of the plasma membrane are continuously inserted from G_1 to G_2 (Graham et al., 1973), but the thickness of the membrane remains constant (Knutton, 1976). Pasternak and colleagues (1974), using mouse P815Y cells in suspension synchronized by zonal centrifugation, found that many of the carbohydrate-containing compounds were predominately synthesized in G_1. The protein of the membrane doubles between G_1 and G_2, and since the phospholipid:protein ratio remained constant, they concluded that phospholipid also was synthesized continuously through G_1 to G_2. The volume of the cell doubles before mitosis, and the surface area increases 1.6-fold. The extra surface area required for cytokinesis is provided from the microvilli.

6.4.1. The Cytoskeleton

Alterations in cell shape are likely to involve interactions between the contractile protein in the microfilaments, transmembrane proteins, and surface proteins such as fibronectins (LETS protein) (see Table 6.2). Fibronectin is a multifunctional glycoprotein which exists on the cell surface and extracellularly. It consists of two subunits of 200 ± 20 kDa linked by a disulfide

TABLE 6.2. PROTEINS OF THE CYTOSKELETON

Protein	Subunit (M_r kDa)	State	Function	Structure	Size (nm)	Inhibitors
Actin	42	Polymer	Contractile	Microfilaments	6	Cytochalasin-B
Myson (H-chain)	200	Dimer	Contractile			
Tropomyosin	35	Dimer	Stabilization of actin bundles			
Actin-binding protein	220	Dimer	Cross-linking			
Filamin	240	Dimer	Polymer in gels			
Gelsolin	91	Dimer	Ca^{2+} regulation of gelation			
Profilin		Monomer	Regulation of actin polymerization			
α-Actinin	102	Monomer	Attachment to membranes			
Vimentin	52	Polymer	Fixes position of nucleus	Intermediate Filaments	10	Colcemid
Keratin	40–65	Polymer	Structural protein of epidermis			
Desmin	55	Polymer	Mechanical integration of myofibril z-discs			
Glial protein	51	Polymer	Structural protein of glial cells			
Unnamed	68	Polymer	Forms neurofilaments			
Tubulin (α and β)	55	Polymer	Structural, vesicle guide	Microtubules	25	Colchicine, vinblastine
Dynein	370	Dimer	ATPase			

Source: Houslay & Stanley, 1982.

bond near the carboxyl end (Hynes & Yamada, 1982). By degradation with different proteases, five or more polypeptides can be released and characterized by their capacities to bind to a range of macromolecules such as actin, collagen, and the plasma membrane (Yamada et al., 1981). Fibronectins are found in plasma, soft connective tissues, and most basement membranes; they are synthesized by many cells including fibroblasts, endothelial cells, chondrocytes, myoblasts, hepatocytes, amniotic cells, and early embryonic tissues (Hynes & Yamada, 1982). Fibronectin promotes cell migration in vitro (Ali & Hynes, 1978). It had been thought to be responsible for adhesion of cells to plastic surfaces, but recently Birchmeier (1981) has shown it is not immediately concerned in making focal contact between microfilament bundles in fibroblasts and the plastic surface.

Microfilament bundles are made up of subunits of actin (42 kDa), generated by ATP-dependent polymerization of G- to F-actin. A miscellany of proteins, the most important of which are filamin and actin-binding protein, serve as cross-linking molecules to form gels. These cross-linkers have been found in numerous cell types, for example, ascites, sea urchin eggs, sperm, and macrophages. Tropomyosin is associated with actin in the microfilaments, and in nonmuscle cells it stabilizes the condensation of the filaments into stress fibers. Stress fibers connect the nucleus to the plasma membrane, and in locomotion they line up with the direction of movement. In lammelipodia the blebs and ruffles contain only a meshwork of actin, which suggests that microfilaments are implicated in maintaining a flattened shape, and the actin meshwork in ruffling and locomotion (Houslay & Stanley, 1982).

Actinin and vinculin (130 kDa) are present in focal contact areas at the end of the microfilament bundles. In cells transformed by oncogenic viruses, the src protein kinase phosphorylates tyrosine residues in vinculin. It is not yet clear if this is linked to the rounding of malignant cells nor if vinculin phosphorylation is important in releasing cell-cell contacts in normal tissues (Sefton et al., 1981; Virtanen et al., 1982).

During the fusion of myoblasts which occurs with cells arrested in G_1, there is a severalfold increase in amount of fibronectin (Teng & Chen, 1975). Blumberg and Robbins (1975) established correlations between the removal of fibronectin from the cell surface and the mitogenic capacity of various proteases. On the other hand, experiments with thrombin (Section 5.2.2) indicated that loss of fibronectin was not a prerequisite for mitogenesis or progression through the cell cycle (Hynes, 1976).

In addition to the just-mentioned cytoskeletal elements there are the so-called intermediate (10 nm) filaments, which are distinguished by their biochemical heterogeneity and relative insolubility under physiological conditions (Lazarides, 1982). Different cell types have different components: epithelial cells contain keratinlike compounds (cytokeratins), others contain vimentin or desmin, and neurofilaments are characteristic of neuronal cells and glial filaments of astrocytes. This generalization is not absolute; Dräger (1983) found neurofilaments and vimentin in the neuron of adult mouse retina, and Kreis and colleagues (1983), by microinjection of poly (A^+)mRNA, isolated from bovine epidermis, into nonepithelial cells containing only intermediate filaments of the vimentin type, showed that keratinlike poly- peptides could be synthesized and be assembled into intermediate filaments. They concluded that the assembly mechanisms exist in vivo which sort out newly synthesized cytokeratins from vimentin. A very exhaustive analysis of human cytokeratins has been provided by Moll and colleagues (1982), who have identified 19 different polypeptides, each probably a unique gene product.

At cell division all the major structural elements are reorganized, as are, for example, the microtubules and actin filaments which are engaged in the mitotic process. On the other hand, the intermediate filaments have a much less well defined function. Immunofluorescent techniques have shown (Lane et al., 1982) that, during epithelial cell division, the filaments are replaced by a "speckled pattern of cytoplasmic dots," but the significance of this is not clear.

One question which arises is the role which cell shape per se, could have on metabolic processes. Folkman and Moscona (1978) noted that bovine endothelial cells adjusted their shape from spheroidal to flat, depending on their adhesiveness to the substratum. The more spheroidal the cells, the lower their incorporation of ^{3}HTdR into DNA. They and others (Bissell et al., 1982) emphasized how conditions affect gene expression. Differentiated rabbit articular chondrocytes were found to produce type II collagen and cartilage-specific proteoglycan. If cells were cultured in monolayer, this was replaced by type I collagen and low amounts of proteoglycan. If the cells were then transferred to firm agarose gels, the flattened cells became spheroidal and reverted to their original phenotypic expression (Benya & Shaffer, 1982).

Shape necessarily affects the surface:volume ratio. We do not know the precise limitations controlling the permissible density of functional molecules/ unit area of membrane surface. Constraints to cell metabolism through restricted transport have already been mentioned. Increased surface area in flattened cells must potentially allow more functional molecules to be present in the plasma membrane. The intimate relationship between the nucleus and the plasma membrane is thus mediated in part via the cytoskeleton in a potentially directional manner. That between the nuclear contents and the cytoplasm involves the nuclear pores. Alterations in the properties of the latter consequent on growth stimulation have still to be elucidated.

6.4.2. Basement Membranes

Basement membranes and the extracellular matrix play a major part in growth and development of regenerating tissue. They are found adjacent to cells which secrete them and are present in almost all tissues of the body. Basement membranes are synthesized by epithelial, endothelial, and smooth-muscle cells (see Farquhar, 1978) but not by fibroblasts or other connective tissue elements. The function of a basement membrane is to support epithelial and endothelial cell layers. It may act as a filtration barrier to molecules moving across capillary walls and epithelial surfaces. The basement membrane is made up of complex dissimilar proteins rich in carbohydrates. The amino acid content of their collagen resembles that of tendon collagen except for the higher proportion of hydroxyproline and hydroxylysine (Kefalides, 1978). Laminin is present only in basement membranes (Foidart & Reddi, 1980); it is made up of three polypeptides, A (450 kDa), B_1 (240 kDa), B_2 (230 kDa), and a fourth, C (150 kDa), which is not a collagen. The polypeptides are linked by disulfide bridges (Cooper et al., 1981). Three antigenic determinants have been distinguished so far, but the precise role which these different sites have in maintaining cell and tissue contacts and separation still has to be elucidated.

The extracellular matrix consists of collagen, glycoproteins, and glycosaminoglycans and is present in tissues which do not have tight junctions between the cells. Excellent accounts of how the extracellular matrix directs gene expression have been given by Bissel and colleagues (1982), and of the importance of tissue interactions in development by Wessells (1977).

The extracellular matrix can have very dramatic effects on morphological redifferentiation. Mammary epithelial cells from virgin, pregnant, or lactating mice resemble fibroblasts when grown on a flat plastic substratum, but when transferred to floating collagen gels, they revert to the pregnant phenotype (Emerman et al., 1981). Demineralized bone collagenous matrix is mitogenic in culture (Rath & Reddi, 1979) and, if implanted below the epidermis, induces the conversion of fibroblasts to chondrocytes and osteocytes, resulting in cartilage and bone deposition. On the other hand, hyaluronic acid, a component of the extracellular matrix, was found to inhibit the synthesis of sulfated glycosaminoglycans (Wiebkin et al., 1975). The proliferative effect of the matrix is exemplified by the maintenance of growth of stratified epidermis. Cells in contact with the basal lamina divide, while those released from their substratum stop dividing, move upward, flatten, and differentiate into the upper stratified layers.

These observations suggest that the development of the blastema into a regenerated limb will be highly dependent on its cellular secretions and its environment. Evidence supporting this comes from Reddi and his colleagues (Gulati et al., 1983). With adult newts in the early stages during the wound-healing process after limb amputation, fibronectin disappears from the muscles of the distal region of the stump. This is followed by a disappearance of laminin. As the blastema develops, fibronectin reappears in the extracellular matrix, which suggests that it is involved in cell adhesion, migration, and alignment of the cells. Laminin is not found in the blastema but may be important in redifferentiation, since it is seen at the site of cartilage differentiation. Blastemal cells lack a basement membrane, but in subsequent development of myotubules laminin appears before fibronectin in the basement region. Since laminin and fibronectin are formed sequentially, the authors suggested that fibronectin is important in the formation of the blastema and in the initiation of redifferentiation, while laminin contributes to the architectural organization of the regenerating limb (Gulati et al., 1983).

6.5. CHEMOTAXIS

In view of the intense reorganization which takes place in the remaining tissue following amputation or severe injury, it is imperative that steps be taken, first to limit the damage, and second to initiate healing before commencing the regenerative process. In order to do this, signals must be sent from the superficial to the deeper regions of the tissue, including those promoting migration of the cells into the area of the wound. Chemotaxis has been thoroughly investigated in the aggregation of the slime mold *Dictyostelium* by cAMP, with pheromones as insect attractants, and in the response of neutrophils, monocytes, and macrophages to bacterial secretions.

Something similar to the last of these could be most relevant in the first stages of regeneration.

Although it had been shown in the last century (Gabritchevsky, 1890) that bacteria or their products were chemotactic for neutrophils, it was not until the 1960s that Keller and Sorkin (1967), and Ward and colleagues (1968) showed that the bacterial attractant was independent of serum. The factor(s) were heat stable, of low molecular weight ($M_r = 0.15-1.5$kDa), with a blocked amino acid terminus and free carboxyl end group, and it was surmised that N-formyl methionyl peptides could be the active substances. Schiffman and colleagues (1978) tested a large number of synthetic peptides in many combinations and permutations. Those with N-formyl methionine followed by two hydrophobic amino acids showed activity at about 10 nM with Leu and phenylalanine (Phe) the most potent. Zigmond (1977) confirmed these observations and further showed that the neutrophils responded readily to a concentration gradient difference across the cell of 10%, with an effect being detectable at the 1% level.

Niedel and Cuatrecasas (1980) reported that Ca^{2+}, Mg^{2+}, and Mn^{2+}, as well as albumin, enhance the binding of the peptides and their chemotactic potency. Initially the receptors are diffusely spread through the membrane, but on binding, they rapidly aggregate and are then endocytosed. The fate of the complex has not been explicitly determined, but inferentially there is an internalization of the receptor, and it is known that the peptide is slowly degraded and released into the medium.

There is no evidence that formylated peptides participate in chemotaxis except as secondary products of prokaryotes, but the peptides Val-Gly-Ser-Glu (Val for valine) and Ala-Gly-Ser-Glu (Ala for alanine) isolated from human lung tissue have been found to be chemotactic in high dilution to eosinophils (Goetzl & Austen, 1975). Human peripheral monocytes demonstrate chemotaxis in response to α-thrombin (a seryl protease) in concentrations around 10 nM, the same range in which the formylated methionyl peptide is active (Bar-Shavit et al., 1983). Pike and Snyderman (1982) have shown that methylation is essential in chemotaxis of mononuclear leukocytes. N-fMet-Leu-Phe stimulated chemotaxis, the release of arachidonate, and superoxide formation, all of which were depressed by agents inhibiting methylation, which was therefore postulated as a requirement for the maintenance of high-affinity state of chemotactic receptors.

Actin has been identified in the submembraneous area at the leading edge of neutrophils responding to N-fMet-Leu-Phe. A release of membrane-bound

Ca^{2+} has been detected by fluorescence studies and electron microscopy has shown deposition of Ca^{2+} in the submembranous regions. This mobilization of Ca^{2+}, together with actin and presumably other contractile proteins, could cooperate in the chemotactic response, but the biochemical basis for chemotactic movement in eukaryotic cells has yet to be provided (see Weatherbee, 1981).

7

THE RESPONSES—
THE CYTOPLASM

We have already seen how signals are released systemically after injury, or locally in response to tissue damage, and how they effect a number of membrane functions directly or consequentially upon the actions of secondary messengers. Cytoplasmic responses (Bresnick, 1971; Bucher & Malt, 1971; Tsanev, 1975; Baserga, 1976; Ord & Stocken, 1980) fall into two operational classes; immediate responses which are apparent within a few minutes and do not require new RNA or protein synthesis, that is, they are unaffected by actinomycin-D or cycloheximide; and longer-term effects which are evident in mammalian cells by 4–6 h after stimulation. The immediate responses include some of those on transport already referred to, as well as allosterically effected promotion of glycogenolysis and other pathways. Their consequence is to remove constraints which are a feature of cells in the G_0 state (Section 1.1.1) and may have prevented the cells from proliferating and may even have been required for them to assume their differentiated functions.

The constraints include:

1. Unfavorable energy charge.
2. Diminished levels of intermediates required for growth (amino acids, nucleoside precursors, etc.).
3. Inadequate cell size. In the case of unstimulated small lymphocytes their very small cytoplasmic volume contains few mitochondria (often < 10), few ribosomes, and little endoplasmic reticulum (see Fig. 7.1).
4. Reduced rates of protein synthesis (see Austin & Kay, 1982).

The immediate changes in the plasma membrane and cytoplasm are followed usually by the hypertrophic response—the cell grows. With some cells there are accompanying marked increases in cell organelles—notably in small lymphocytes (see Fig. 7.1). Some 12–24 h after the stimulus, the hyperplastic phase follows. Cell cycle kinetics and cdc mutants (Chapter 1) have indicated a number of restriction points which cells must pass for replication, mitosis, and cytokinesis to be possible. Their biochemical identification has hardly started. In many cell types in G_0, some obvious requirements for replication are missing or inadequate in amount, notably deoxynucleosides. Regenerating liver and stimulated lymphocytes both show a series of enzyme inductions which cause the necessary intermediates to be produced late in G_1 phase. It is, however, quite clear that the enzymes

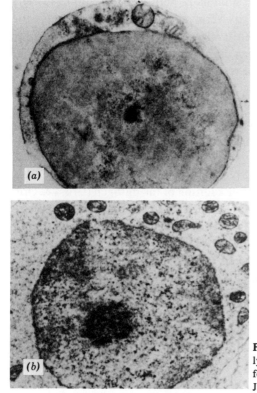

Figure 7.1. Quiescent and transformed lymphocytes. (*a*) Quiescent. (*b*) Transformed. (Photographs courtesy of Sir James Gowans.)

so far identified for these syntheses are neither the sole nor the critical limitations in replication (see Baserga, 1976). With cells that undergo more than one round of division after promotion from G_0, amounts of the induced enzymes frequently remain elevated through the second cycle, but DNA synthesis remains periodic.

7.1. IMMEDIATE EFFECTS INDEPENDENT OF RNA SYNTHESIS

7.1.1. Energy Charge

Intracellular concentrations of ATP in different cell types are usually 5–8 mM (for different muscles see Beis & Newsholme, 1975). Careful studies with what were considered to be minimally perturbed synchronized cells have shown that the energy charge (Atkinson, 1968) can vary through the cell cycle (see Lloyd et al., 1982). It is thought, however, that usually the changes in ATP follow, rather than cause, altered biosynthetic activity.

The idea that cells in G_0 might be restricted in their activities by unfavorable ATP:AMP ratios is one which receives influential support from many metabolic studies on feedback control in vitro, for example, phosphofructokinase, but for which there is little evidence either way in vivo. Accumulated data from this laboratory show unchanged ATP:AMP ratios of about 5 for normal rat liver, for rat liver 24 h after partial hepatectomy, and for random populations of rat thymocytes. In pig lymphocyte cultures stimulated by PHA, the ATP pool rose sharply 0.5 h after addition of the mitogen (Cross & Ord, 1971), but these cells are exceptional because of the very small amounts of their cytoplasm. In neurons it is believed that requirements for ion pumping preempt at least 50% of the available ATP, and similar conclusions have been reached recently for ascites cells (Skog et al., 1982). Generally, however, we have insufficient detailed knowledge from which to draw up intracellular ATP budgets. Binding constants and turnover of the different ATP-dependent reactions are incompletely known, and little quantitative evidence is available about differences in, and potential limitations to, ATP cycling rates between resting and growth-promoted cells. Classical studies on the Pasteur effect established the interdependence between the energy needs of the cell and its glucose uptake; it seems possible that the pleiotropic effects (Hershko et al., 1971) of the growth-promoting agents already discussed may include increased ATP turnover.

7.1.2. Precursor Availability

Amino Acids

Immediately increased transport of amino acids and, later, actinomycin-sensitive increases in amino acid transporters are evident in liver and other growth-promoted tissues (see Tata, 1970 and Chapter 6). In the liver all the blood from the portal outflow has to pass into the residual lobes following partial hepatectomy, and rapid changes in plasma and intracellular amino acid levels are evident (1.5–3-fold increases at 1.5 h) (Ferris & Clark, 1972; Ord & Stocken, 1972a), affecting principally ornithine, lysine, methionine, and the aromatic amino acids.

Intracellular concentrations of amino acids in extrahepatic tissues other than glutamate and aspartate are normally low. The liver plays a central part in the provision of precursors for growth in many tissues, directly as in the synthesis of purine and pyrimidine ribosides, and indirectly because of its role in protein turnover, allowing limiting amino acids to be released into the circulation. There is extensive clinical evidence in humans for the importance of liver and muscle tissue in adjusting and maintaining nitrogen balance after injury (Jeejeebhoy, 1981; Waterlow & Jackson, 1981), but it is not known if similar metabolic alterations occur in Amphibia and invertebrates.

CHOLESTEROL

The need of many cells for cholesterol for growth and proliferation is well established but still unexplained—plasma membrane formation, membrane fluidity, or the need for dolichol-phosphate for the synthesis of glycosylated membrane constituents are all possibilities (Chen et al., 1975; Chen, 1981). Although most vertebrate cells were thought to be able to synthesize cholesterol, it is currently supposed mammalian extrahepatic tissues obtain the bulk of their requirements from the circulation, utilizing circulating low-density lipoproteins (LDL) via LDL receptors. Feedback control over the number of these receptors has been reported in human fibroblasts (see Goldstein & Brown, 1977). It seems likely that, during growth and tissue reconstruction, related controls would ensure cholesterol uptake by fibroblasts and other cells.

Detailed studies (Quesney-Huneeus et al., 1983) with synchronized baby Syrian hamster kidney-E (BHK) cells have established that the cholesterol requirement is a feature of cells in G_1 phase and that, as the cells move into S, amounts of HMG-CoA reductase increase five- to tenfold. If this enzyme was competitively inhibited by compactin, DNA synthesis was blocked, not because of the need for mevalonic acid for cholesterol biosynthesis, but because it is also a precursor of isopentenyladenine, which appears to be specifically required for replication, though not for DNA repair.

NUCLEOSIDES

Nucleosides become available to cells from two sources, exogenously from the extracellular compartment (the *salvage pathway*, Plageman & Wohlhueter, 1980) and by endogenous synthesis *de novo*. Studies on the availability of purine and pyrimidine ribosides during growth were made initially with liver and lymphocytes; generally confirmatory results are becoming available from other tissues. One of the immediate responses to growth-promoting factors in many cell types is increased uptake of pyrimidine ribosides (Section 6.1.3). This does not require protein synthesis and affects translocation, rather than the subsequent phosphorylation step, to give nucleoside monophosphate. With liver, ribonucleoside and ribonucleotide pools are almost unaltered by the potentially increased uptake for the first 5 h after partial hepatectomy. The most important practical consequence of the increased uptake is the substantially greater labelling of the precursor pool (see Bucher & Malt, 1971; Ord & Stocken, 1973a), which complicates interpretation of radioisotope experiments on RNA metabolism in growth-promoted cells. By 6 h the level of uridine nucleotides in regenerating liver is increased, especially in respect of uridine diphosphate sugars required for glycosylation of membrane components (see Lesch et al., 1973); increases in cytidine triphosphate (CTP) appear later (12–24 h) (Bucher & Malt, 1971; Jackson et al., 1980). How these changes are achieved is not certain but may be only the establishment of a new steady state, following the increased capacity for translocation and the increased removal of uridine into compounds outside the soluble nucleotide pool.

In thymocytes and peripheral lymphocytes, salvage and endogenous pathways are both available, and both are promoted after mitogen stimulation (Ito & Uchino, 1973; Ling & Kay, 1975; Thuiller et al., 1982), the endogenous pathway increasing markedly as DNA synthesis occurs. The endogenous pathways appear susceptible to feedback inhibition from the salvage pathway if exogenous nucleosides are available.

Compartmentation of ribonucleotide pools has been detected, but in most cell types examined equilibrium between nuclear and cytoplasmic compartments is very rapid (Khym et al., 1979).

Deoxyribonucleotides become available in cells as they approach S phase. This will therefore be discussed later.

7.1.3. Translational Control of Protein Synthesis

Enhanced protein synthesis as cells move out of G_0 could arise for four reasons:

1. Increased translation of preexisting messenger RNA.
2. Increased posttranscriptional processing of RNA.

3. Increased transcription

4. Initiation of new transcription.

Probably all of these occur, but the first is seen in regenerating liver and lymphocytes soon after stimulation and can be independent of transcriptional changes (Jagus & Kay, 1979; Austin & Kay, 1982).

Translational control in eukaryotic cells has been studied classically in *Acetabularia* (Hämmerling, 1953, 1963) and in vertebrates in reticulocyte lysates. In the latter, synthesis of globins is strongly dependent on the presence of heme. In its absence protein synthesis is reduced to about 10% of its initial rate because of the activation of an inhibitor which phosphorylates and inactivates the subunit of initiation factor eIF-2. There is not a quantitative relation between the phosphorylation and the inactivation. It is likely that further interactions which occur during the formation of the 40 S mRNA initiation complex, entailing GTP binding and the subsequent dissociation and breakdown of eIF-2-GDP, are retarded if eIF-2 is phosphorylated (Austin & Kay, 1982; Clemens et al., 1982; Pain & Clemens, 1983).

Translational inhibitors which phosphorylate eIF-2 have now been found in many cell types including liver and lymphocytes. Formation of initiation complexes is further influenced by the structure of the mRNAs, the more rapid initiation by β- than α-globin mRNA being an obvious example. Rates of elongation of different peptides are relatively constant. Factors regulating the formation of initiation complexes can therefore alter the pattern of protein synthesis at the translational level; whether this is important in growth is not known.

In pig lymphocytes stimulated by PHA, Jagus and Kay (1979), using ^{35}S-met tRNA$_f$, showed that the capacity to form initiation complexes was enhanced. In 2–20 h after the addition of the mitogen, the increased rate of protein synthesis was directly proportional to the increased mRNA binding to the ribosome. Increased initiation was also found in growth-promoted fibroblasts (Meedel & Levine, 1978) and hepatoma cells (Streumer-Svobodova et al., 1982).

We have already discussed increased phosphorylations following growth-promoting signals. It is not known if changes in phosphorylation of eIF-2 occur, but the effect would be in a direction contrary to the change in protein synthesis observed (and perhaps better correlated to the "switched-down" protein synthesis found in cells at confluence, with their high cAMP levels). How dephosphorylation of eIF-2P is achieved, and its potential control, are not known.

Phosphorylation of ribosomal proteins could also occur and has been reported from ^{32}P studies after partial hepatectomy (see references in Ringer et al., 1981). Interpretational difficulties here were considered earlier (Section 5.2.5) and are compounded by results of detailed measurements of the ^{31}P-phosphate content of ribosomal proteins obtained from normal and partially hepatectomized rats. These showed increased phosphate content from the membrane-bound 40 S ribosomal subunits along with decreased ^{31}P in the proteins from the 60 S subunit. Phosphate contents from free ribosomes were unchanged (Ringer et al., 1981).

Other regulatory influences on translation in eukaryotes are known (see Austin & Kay, 1982). ATP availability and redox state are only relevant in some cell types, but increases in intracellular amino acid levels may be of greater significance. In prokaryotes an increase in the ratio of uncharged to charged amino acyl-tRNA is inhibitory. This is not found with eukaryotes, but a fall in the supply of amino acids still causes an immediate drop in protein synthesis (Austin & Clemens, 1981). In Ehrlich ascites cells deprived of lysine, this effect was again at the level of initiation—amounts of met-tRNA$_f$ complexes were lowered—and was reversed by adding eIF-2, but there was no direct proportionality between the reduction in lysine concentration and the level of phosphorylated eIF-2. How amino acid availability regulates initiation of translation in eukaryotes is still unknown.

Translational control continues to be important when increased transcription is also occurring. Cooper and Braverman (1981) demonstrated that amounts of initiating met-tRNA$_f$ were rate limiting in the formation of the 40 S initiation complex in human lymphocytes: direct proportionality was found between levels of met-tRNA$_f$ and the rates of protein synthesis. By 5 h after addition of PHA, amounts of tRNA$_f$ were increased twofold, and by 20 h the increase was fivefold. No change in the proportion of charged (amino acylated) tRNA was detected (80%).

7.2. THE HYPERTROPHIC PHASE

7.2.1. Increased Efficiency of RNA Processing

RIBOSOMAL RNA

Classical studies demonstrated a net increase in cytoplasmic rRNAs in regenerating liver by about 6 h after partial hepatectomy and a little later in stimulated lymphocytes. Detailed studies in rat livers 12 h after partial hepatectomy (Dabeva & Dudov, 1982; Dudov & Dabeva, 1983) have established that this was largely due to increases in the efficiency with which pre-rRNA in the nucleolus was processed and translocated to the cytoplasm. The half-life of all species of nucleolar pre-rRNA and rRNA was reduced by an average of 30%. This was associated with a shift in maturation route of pre-rRNA to favor the most direct route: 45 S→32 S and 18 S; 32 S→28 S RNA. As the result of the changes, about three times more ribosomes were synthesized and appeared in the cytoplasm, thus making a major contribution to increased levels of protein synthesis. Maturation of pre-rRNA occurs by endonucleolytic cleavage. Why the pattern and/or rates of hydrolyses should alter has not yet been established. Dudov and Dabeva (1983) suggest that activated ribosomal protein synthesis, which has been reported a few hours after partial hepatectomy (Nabeshima & Ogata, 1980), allows faster assembly of pre-rRNA particles which are needed for maturation.

MRNA

Originally it had been expected that movements of cells from G_0 into an active growth cycle would entail the transcription of messages not expressed in cells in G_0. In the 1950s increased precursor uptake into total cellular RNA was found for many growing tissues, and later it was found that their growth was inhibited by actinomycin-D. As our understanding of RNA metabolism has advanced, so expectations of likely changes in growth have shifted. When it became clear that more mRNA was transcribed in the nucleus than reached the cytoplasm (see Harris, 1974), the latter now being estimated at 1–2% of the RNA transcribed (Tobin, 1979), attention turned to the actinomycin-sensitive, rapidly turning-over, heterogeneous (Hn) fraction of nuclear RNA. The discovery of pre-mRNA and its processing, especially polyadenylation at the 3′ end, offered a convenient experimental procedure for separating most nuclear mRNA precursors and polysomal mRNA, excepting the minor amounts of poly (A)-free small RNAs, which are more difficult to analyze. Currently, processing pre-mRNA (5′ capping, splicing and polyadenylation) is under consideration as a set of important, regulated stages which could influence the amounts of mRNA available for translation independent of changes in primary transcripts. We have already seen how processing changes are important for rRNA; possible alterations for pre-tRNA have not yet been extensively examined.

Most of the work on mRNA changes in growth has been performed with regenerating liver because of the ready availability of material. Very young rats have been used to increase synchrony and thus the precision of the results. Hybridization procedures have become increasingly sophisticated theoretically and experimentally (Wetmur, 1976); for this reason, only rather recent results are considered, where most groups are agreed that:

1. Amounts of polysomal polyadenylated mRNA are more than doubled in the first 24 h after partial hepatectomy (Atryzek & Fausto, 1979).

2. The sequence complexity (range of RNA molecules transcribed) of nuclear RNA and polysomal poly(A)$^+$ mRNA at 12 h (hypertrophic phase) and 24 h (hyperplastic phase) is not detectably different from that in sham-operated rats (Tedeschi et al., 1978; Scholla et al., 1980; Grady et al., 1981).

3. It is likely that the relative proportions of the different types of polysomal mRNA change (Scholla et al., 1980; Wilkes & Birnie, 1981). Essentially identical results to these have been reported from conA stimulated bovine lymphocytes (Kecskemethy & Schäfer 1982). Poly(A)$^+$ mRNA increased threefold after 40 h stimulation. If the mRNA was translated in vitro, actin, tubulin, and calmodulin could be identified in the products, in different proportions from those in quiescent cells. All those in the field emphasize that the hybridization procedures cannot detect small ($< 0.1\%$) changes in transcription from the least abundant, nonrepetitive DNA.

No information is available about how the increased amount of polysomal poly(A)$^+$ mRNA appears, but since the process affects all constitutively expressed mRNA, it seems likely that nonspecific enhancement of one or more processing stages is involved. From our earlier considerations of secondary messengers (Chapter 5), increased activity of processing enzyme(s) through changes in protein modification requires examination.

Complementary to the statistical analysis of all RNA molecules in subclasses of differing hybridization kinetics is the assay of amounts of individual types of mRNA specifying particular proteins. If the mRNA can be isolated, an estimation is now possible, following synthesis and cloning of the cDNA. Although this is now in hand for some liver proteins (tyrosine aminotransferase, EC 2.6.1.5., Diesterhaft et al., 1980), it is only just starting to be used in studies on growth.

7.2.2. RNA Polymerases

Increased activity of RNA polymerases during growth had been observed before the complexities in eukaryotic RNA polymerases were apparent.

With the availability of techniques by which the three polymerases can be distinguished and separated, it is clear that all three DNA-dependent RNA polymerases in rat liver are increased in activity at least by 18 h after partial hepatectomy. RNA polymerase (pol) I activity (rRNA transcription) goes up fourfold, and RNA pol II (mRNA) and RNA pol III (tRNA and other small RNA species) are increased twofold (Duceman & Jacob, 1980). Immunochemical or other data to establish whether the changes are due to increased amounts and/or activities of the enzymes are still not available.

7.2.3. New Transcription

Present limitations in deciding the extent to which increased preexisting primary transcription (rates of initiation of pre-mRNA and/or its elongation), or the promotion of new transcription, occur during growth have already been mentioned. It is apparent from the analysis of increased polysomal poly(A)$^+$ mRNA that most of the newly synthesized protein is constitutive and already in quiescent cells. This is supported by detailed data from cell cultures and yeasts showing that amounts of protein increase continuously through the cell cycle. It is very difficult from kinetic studies to decide whether the increase in protein is linear, exponential, or stepped, but for a small number of specific proteins there is clear evidence of periodic synthesis (see John et al., 1981; Lloyd et al., 1982). These include ornithine decarboxylase (EC 4.1.1.17, Russell & Haddox, 1979), a number of enzymes concerned with deoxynucleoside metabolism, and histones. Ornithine decarboxylase shows complex variations in activity through the cell cycle (see Ornithine Decarboxylase, following), but alterations in deoxynucleoside metabolism in some cells are associated with a restriction point after which the cells are committed to replication. Net histone synthesis occurs in S phase.

ORNITHINE DECARBOXYLASE

Ornithine decarboxylase is the rate-limiting enzyme in polyamine synthesis in many cells (see Fig. 7.2); in these cells its activity is increased markedly in G_1 when growth is promoted (see Russell & Haddox, 1979). In mammalian liver it is one of the enzymes that shows diurnal variations in activity (Hopkins et al., 1973), and its response to partial hepatectomy is dramatic and complex with at least two peaks in activity (see Fig. 7.3), probably resulting from interactions between the induction due to tissue depletion and from normal response to feeding patterns (Hölta & Jänne, 1972; Thrower & Ord, 1974).

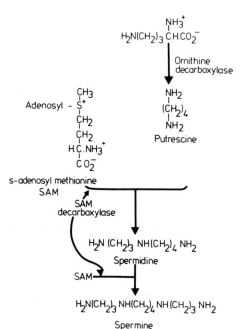

Figure 7.2. Pathways in polyamine biosynthesis.

The enzyme is induced in the different cell types by many of the signals we have already seen to be growth promoting (Chapter 4). In liver, growth hormone, glucocorticoids (see Russell & Durie, 1978), and adenylate-cyclase-stimulating agents are effective inducers (Beck et al., 1972).

Ornithine decarboxylase can be inhibited by substituted ornithines and 1,3-diaminopropane (see Russell & Haddox, 1979). All these agents so far tested prevent cell replication. The function of the small, positively charged polyamines is still unclear (Tabor & Tabor, 1976; Cohen, 1978; Russell & Durie, 1978; Goyns, 1982). Because of their charge they have marked effects on nucleic acid structure and some enzyme activities. Polyamines can be conjugated to proteins, and increased activity of a nuclear transglutaminase has been reported 4 and 42 h after partial hepatectomy (Haddox & Russell, 1981). With the finding that casein kinase II (nuclear protein kinase II) can be activated by polyamines (Section 5.2.4), a messenger role for polyamines has also been advanced. These more recently suggested roles have not yet been tested in inhibitor studies. The parallel (with slight temporal precedence) of ornithine decarboxylase induction with an increased RNA polymerase I activity suggested that ornithine decarboxylase might be an initiation factor for RNA pol I in nucleoli (see Russell & Durie, 1978). When highly purified ornithine decarboxylase was added to nuclei, RNA pol I activity was increased; this response was partially prevented by ornithine decarboxylase inhibitors and was not mimicked by putrescine alone.

Ornithine decarboxylase has one of the shortest half-lives for mammalian enzymes ($t_{1/2} = 10-11$ min). It was established immunochemically that the increase in activity in growth-promoted cells was due principally to increased amounts of enzyme; antagonism from ornithine decarboxylase antienzyme has already been considered (Section 5.2.4). The rate of destruction of the enzyme is not thought to be altered in a major way during growth (see Russell & Durie, 1978). The enzyme may, however, be inactivated by a transglutaminase catalyzed conjugation to putrescine, an end product of ornithine decarboxylase-dependent reaction (Russell, 1983).

As far as we are aware, mRNA and cDNA for ornithine decarboxylase have not yet been used to determine how many copies of the gene are present or whether the increased amount of enzyme is due to increased initiation by RNA pol II or more efficient processing or translocation of its pre-mRNA. The induction of the enzyme is sensitive to actinomycin-D for brief periods after the stimulus is given (> 0.5 h after partial hepatectomy in rats).

One further general point must be raised in respect to enzyme induction in growth-promoted animal cells. What are the inducers? No inducers have

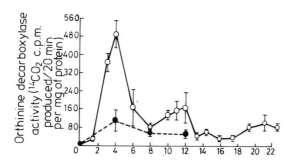

Time of death after operation (h)

o—o—o partially hepatectomised; •—•—• sham operated

Figure 7.3.   Ornithine decarboxylase levels in regenerating rat liver (Thrower & Ord). Reprinted by permission of the Biochemical Society (*Biochemical Journal*, Vol. 144, p. 365, 1974).

yet been identified in adult tissues in the sense familiar in prokaryotic biochemistry, that is, a ligand binding specifically to a protein which itself binds specifically to a unique region of the DNA and elicits the transcription of one or a small number of structural genes.

Russell and her colleagues (see Russell & Durie, 1978) have summarized the extensive evidence implicating cAMP as the secondary messenger eliciting ornithine decarboxylase induction. They further suggest that the specific, critical factor is the increased activity of type I cAMP-dependent protein kinase. In weanling rats, when ornithine decarboxylase was induced by growth hormone, increased cAMP-dependent protein kinase activity was detected, although cAMP levels were not significantly altered. It is not clear whether similar increases in kinase activity would be found after steroid administration and induction; usually such activation is not detected in liver after dexamethasone administration. If cAMP-dependent protein kinase was the only inducing signal, it is also uncertain why adrenergic blockers, which did not reduce cAMP levels in regenerating livers 4 h after partial hepatectomy, did diminish ornithine decarboxylase induction at this time (Thrower & Ord, 1974).

In the early 1970s it was expected that an amino acid or its metabolite would be the inducer in liver for ornithine decarboxylase, both in growth promotion and for the nutritionally linked, diurnal increase (see Fausto, 1971). We have already seen that elevated intracellular amino acid levels are an inevitable consequence of growth-promoting signals. Nevertheless it seems likely that the inductions in adult cells moving out of G_0 into growth are more similar to promotion in prokaryotes, with the steroid machinery or possibly eukaryotic cAMP protein kinase regulatory protein(s) as promoter site-binding analogues. Less-specific effects of cAMP through phosphorylation of nonhistone proteins are also possible. In summary, ornithine decarboxylase induction in mammalian cells is effected through a cAMP-dependent mechanism, as well as by steroids and other growth-promoting factors (Mufson et al., 1977).

In the parallel case of tyrosine amino transferase induction in liver, classical studies by Tomkins in the 1960s with the hepatoma cells showed additivity between the inducing effects of cAMP, insulin, and glucocorticoid. It has also been established for tyrosine aminotransferase that inducibility is not possible throughout the whole cell cycle. With hepatoma cells synchronized by mitotic shaking or serum deprivation, dexamethasone was effective only in the later two-thirds of G_1 phase or in S, although no changes in steroid uptake or binding were detected (Martin et al., 1969; van Wijk et al., 1979). Chromatin changes in the cell cycle therefore influence sensitivity to, and may be a necessary prerequisite for, steroid regulation. The possible involvement of phosphorylation by nuclear kinase I in these inductions has been indicated from its sensitivity to Be^{2+} inhibition (Ord & Stocken, 1981a; Cummings et al., 1982).

ENZYMES IN DEOXYNUCLEOSIDE METABOLISM (FIG. 7.4)

In quiescent cells, excluding liver parenchymal cells, amounts of deoxynucleoside are very low. As the cells pass through their first growth cycle after stimulation, deoxynucleoside metabolism increases (see Rothstein, 1982). Most cells can utilize both exogenous and endogenous pathways to obtain DNA precursors, and in many cells the first sign that they have become committed to replication is the uptake of ^{3}HTdR. Phosphorylation by thymidine kinase is required to trap thymidine intracellularly. The enzyme is present in only very small amounts in, for example, resting lymphocytes, in G_0, and in early G_1, but its synthesis is increased after about 12 h stimulation in lymphocytes (see Barlow & Ord, 1975; Tyrsted & Munch-Petersen, 1977). Increased endogenous synthesis in lymphocytes occurs a few hours later, nearer the start of S phase (Barlow, 1976).

The formation of deoxynucleotides in the liver or liver derived cells requires more complex changes than in the other cells because of the essential role of liver parenchymal cells in the catabolism of thymidine to β-

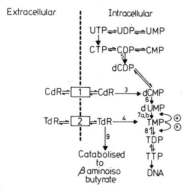

Figure 7.4. Thymidine metabolism in liver (see Stryer, 1981). (1, 2) Pyrimidine deoxynucleoside transporters; (3, 4) Pyrimidine deoxynucleoside kinases; (4) Thymidine kinase (EC 2.7.1.75); (5) Ribonucleoside diphosphate reductase (EC 1.17.4.1); (6) Deoxycytidylic acid deaminase; (7a, b) Thymidylate synthetase (EC 2.1.1.45) Dihydrofolate reductase (EC 1.5.1.3); (8) Thymidylate kinase; (9) Hydrolysis and deamination.

aminoisobutyric acid. Unlike extrahepatic tissues, thymidine is taken up by resting liver easily, virtually complete oxidation following facilitated transport. Like uridine but not cytidine, thymidine entry is increased immediately after partial hepatectomy (Ord & Stocken, 1972a, 1973a), the affinity of the transporter for the ligand being increased (Ord & Stocken, 1973b). TTP is obligatorily required for DNA synthesis whether it is formed from exogenous, salvaged thymidine or *de novo* (see Fig. 7.4). Reduction in its catabolism is therefore essential (Canellakis et al., 1959). These opposing metabolic requirements illustrate a biochemical dilemma which differentiated cells may encounter as they move out of the G_0 state. In this case it is solved by a set of changes in enzyme activities which are thought to be due primarily to their syntheses, reinforced by allosteric feedback controls of thymidylic acid metabolism (see Fig. 7.4).

Thymidine catabolism was found to decline from 16 to 20–22 h after partial hepatectomy, and thymidine kinase activity rose, so at the peak of thymidine incorporation into DNA, amounts of β-aminoisobutyric acid/g liver wet weight were only 50% of those found at 16 h, while levels of thymidine nucleotides increased threefold (Söderhäll et al., 1973; Ord & Stocken, 1975a).

These effects and the increased activities of ribonucleotide reductase (see Thelander & Reichard, 1979), dihydrofolate reductase (Mariani et al., 1981), and deoxycytidylate deaminase appear to be coordinated. Agents which block one reaction, such as α-adrenergic blockers in liver or exposure to X-irradiation, affect the others also.

Two further developments in deoxynucleoside metabolism may ultimately shed light on the controls by which the machinery is switched on at the period in G_1 phase when cells are committed to replication (or marking the end of G_0). The first is the development of DNA transfer procedures whereby the gene for thymidine kinase (tk^+) can be introduced into tk^- cells, for example, mouse LMTK$^-$ cells (Schlosser et al., 1981). Potentially this should enable transcriptional controls for the tk gene, and presumptively others in the same coordination set, to be identified.

The second observation is that a number of enzymes involved in the biosynthesis of deoxynucleotides may become associated as a multienzyme complex in nuclei as the cells go into S phase (Reddy, 1982; Reddy & Pardee, 1982). The possible implications of this and further aspects of structural changes in nuclei as the cells go into S are reviewed in Section 8.4.

7.2.4. Decreased Protein Catabolism

A reduction in rates of protein degradation as the basis or accompaniment of growth has been a logical possibility since protein turnover was discovered. That cells could regulate their rates of protein catabolism was first found by Mandelstam (1960) in studies with *E. coli*. Since then, increasing numbers of eukaryotic systems have been found to show the same behavior (see Goldberg & St. John, 1976; Ballard & Francis, 1983; Gunn et al., 1983). The extent of the reduction in protein breakdown can be very appreciable. In regenerating liver 36 h after operation, rates of degradation are halved (Scornik & Botbol, 1976). The diminished breakdown affects proteins at least in the plasma membrane and soluble fraction of the cell; specific differences between individual proteins were not detected (Tauber & Reutter, 1978). Lowered breakdown was also shown in hypertrophic kidney and muscle (see Goldberg & St. John, 1976). In humans after major surgery, the reverse situation is evoked; both protein synthesis and catabolism are accelerated, and if recovery does not ensue, rates of breakdown increase substantially (Munro, 1979; Clague, 1982).

Because of the importance of liver and muscle in protein metabolism in higher animals, it was originally thought that altered protein breakdown might be restricted to these tissues, insulin and glucocorticoids being particularly implicated as likely regulators. Work with cultures has now shown that other cell types respond similarly to liver and muscle, reducing intracellular protein breakdown when growth is promoted (Hopgood et al., 1981). With CHO cells it may be amounts of uncharged tRNAs which are "sensed" (Scornik, 1983). Phe, tryptophan (Trp), or met tRNA might be critical factors in liver.

Hershko and colleagues in 1971 suggested reduced protein breakdown was part of the pleiotropic response to growth-promoting agents. While there is little doubt that this is true, we have no idea how it is accomplished. There are two classes of proteolysis — specific and nonspecific. Hydrolysis at unique sites in particular proteins by specific proteases may allow less-specific hydrolases to act. Such results as are at present available suggest it is the nonspecific proteolysis of constitutive protein which is regulated in growth. Growth-associated changes in rates of breakdown of specific, periodically expressed proteins (ornithine decarboxylase, the deoxynucleoside set of enzymes, histones) have not so far been detected.

Intracellular protein breakdown is an active process requiring ATP and protein synthesis. Lysosomal enzymes are certainly implicated, but there is also nonlysosomal breakdown (see Goldberg & St. John 1976; Tanaka et al., 1983) which may be especially important at least for some inducible enzymes (e.g., see Grinde & Johnson, 1982). The factors which normally influence turnover rates for proteins are slowly emerging, for example, size, presence or absence of prosthetic groups, hydrophobicity, but these are irrelevant to alterations in rates of breakdown in vivo. The role of the lysosome and autophagic digestion is still controversial (see Holzer & Heinrich, 1980), but an interesting possibility being investigated (see Ballard & Francis, 1983; Gunn et al., 1983) is that the rates of autophagic sequestration and fusion with the lysosomes may be variable. Growth promotion affects Na^+-H^+ exchange in cells (Section 5.2.2). A slight rise in intracellular pH might reduce lysosomal fragility. NH_3 can also accumulate in lysosomes in liver and muscle in vitro and inhibit cathepsins through the increase in pH (see Holzer & Heinrich 1980).

7.3. NEW ORGANELLE FORMATION

Cell growth necessitates the formation of new organelles, but the manner by which this is regulated is not known. In some highly differentiated cells like B lymphocytes, increasing numbers of mitochondria, lysosomes, and amounts of endoplasmic reticulum are clearly evident 24 h after stimulation, before significant replication has occurred (see Fig. 7.1). As we have already seen, this is preceded by the formation of increased numbers of ribosomes and accompanied by an increase in cell size and therefore in area of the plasma membrane. Ultrastructural changes in regenerating hepatocytes have been discussed (see Grisham et al., 1975; Murray et al., 1981), but little data is available for the behavior of organelles in the growth cycle of higher animals (Lloyd et al., 1982). Nuclear changes in growth are considered in Chapter 8.

7.4. THE G_1/S RESTRICTION POINT

For a cell to move from growth to division—hypertrophy to hyperplasia—it has to pass through a number of restriction points (Section 1.1.2). We have already considered the biochemical differences between cells in G_0 phase and those moving into G_1. Now the little biochemical information which is available to identify the G_1/S restriction point will be reviewed.

The genetic approach being used most extensively so far with yeast has not yet identified the functions for which the cdc-28 gene product is critical (Section 1.1.2). As will now be apparent, the execution point for a particular gene product (the time of its actual participation in cellular events) need not be directly related to the time of transcription of the gene sequence. In the most extreme case continuous transcription of pre-mRNA can be occurring, but the mRNA may only bind onto the ribosomes and get translated during a limited period (cf. Section 7.1.3). Even greater temporal displacement can occur if the protein is synthesized in a precursor or otherwise inactive form.

The classical biochemical approach by which to identify critical processes at the restriction point has been to use inhibitors which blocked the onset of replication. These inhibitors are ineffective if administered after the cells have passed the restriction point, although replication has still not commenced. Analysis of the changed levels of intermediates can then indicate which pathways are involved.

Many of the analyses are from regenerating liver. Biochemical investigations of this sort are not very far advanced with yeast. Although with eukaryotic cell cultures homogeneity in cell type and minimal perturbation synchrony are better, biochemical data are not extensive, partly because of paucity in material for analysis. Homokaryotic nuclear transplantation and cell fusion have the same drawbacks, although demonstrating clearly alterations in the biochemistry of the cytoplasm between G_1 and S phase cells. A further difficulty is the inadequacy of our understanding of the biochemistry of eukaryotic replication and its initiation.

In regenerating liver, preferably in young rats with their environment and nutrition closely controlled to entrain diurnal rhythms and regenerative events (Section 3.3.2), some biochemical markers can be identified and ordered around the restriction point 12–15 h after partial hepatectomy (see Table 7.1). A number of agents interfere with the orderly sequence of appearance of these markers (see Table 7.2) and cause replication to be delayed by 3–6 h after the normal peak of ³HTdR incorporation into DNA, 22–24 h after operation (see Holmes, 1956; Thrower & Ord, 1974). Usually the amounts of blockers administered have produced delayed, rather than totally inhibited, replication to minimize the production of toxic side effects.

TABLE 7.1.   BIOCHEMICAL EVENTS ASSOCIATED WITH TRANSIT
THROUGH THE RESTRICTION POINT IN RAT LIVER 12–15 HOURS
AFTER PARTIAL HEPATECTOMY

Time in Hours after Operation		Marker	Signal
12–13	1	Peak in cyclic AMP	Adrenaline/glucagon
	2	Peak in ornithine decarboxylase	Glucocorticoids/ somatomedins?
	3	Amount CTP increases	
	4	Increased activity enzymes yielding deoxynucleoside triphosphates	
12–15	5	Peak in phosphorylation of histone H1, initially on "old" molecules	
	6	Phosphorylation of old and new histones H2A and H4	
15–18	7	Decreased activity of thymidine catabolizing enzymes	
	8	Histone synthesis begins	
	9	Thymidine uptake into DNA detectable	

Note: See Text and Ord & Stocken, 1980.

Studies with many of the blockers have established that the lowering in activities they produce when administered in vivo are not seen if the agents are tested against the enzymes in vitro. X-irradiation and dimethylnitrosamine are also much less affective in diminishing DNA synthesis if given when the cells have moved into S phase.

Three features emerge from this analysis:

1.   The chemical diversity of the blockers.
2.   The difference in times at which they must be administered to produce an effect.
3.   The common end effect produced—DNA synthesis is delayed.

Three actinomycin-D-sensitive periods are known for rat livers after partial hepatectomy (see Thrower & Ord, 1974): the first for 6 h after operation, as the cells move out of G_0 into G_1 phase and increasing amounts of all species of cytoplasmic RNA are appearing (see Section 7.2); the second 10–11 h after operation; and a third period around 18 h, at the start of S phase. If given in any of these periods, actinomycin will reduce the extent of DNA synthesis at 21–22 h. Pindolol and phenoxybenzamine, given at the time of operation, affect early G_0/G_1 markers (Section 7.2), but if their administration is delayed until 9–12 h, the first series of events will be completed normally. Be^{2+} and sublethal X-irradiation given immediately after operation do not detectably affect RNA synthesis nor the other early G_0/G_1 changes, but they still retard transit through the G_1/S restriction point.

TABLE 7.2.   AGENTS WHICH ARREST PROGRESS THROUGH THE G₁/S RESTRICTION POINT IN REGENERATING RAT LIVER

Agent	Time of Its Administration in Hours after Operation	Marker Affected	Basis of Action	Reference
Be^{2+}	0–3	4,6,9	Enzyme induction blocked, casein (nuclear) kinase I inhibited	Witschi, 1970; Kaser et al., 1980; Cummings et al., 1982
Sublethal X-irradiation	0–12	4,5,7,9 NOT 1,2,6	Double-stranded breaks in DNA, oxidation thiol groups	Holmes, 1956; Ord & Stocken, 1968
Dimethyl-nitrosamine	6	9	Alkylation of DNA especially 7-methyl guanine	Craddock, 1975
Actinomycin-D	10	8,9 NOT 1	Transcription blocked	Thrower & Ord, 1974
Phenoxy-benzamine	9–12	2,5,8,9 NOT 1	α-receptors blocked, nuclear histone H1 kinase induction blocked	Thrower & Ord, 1974; Thrower et al., 1973
Pindolol	12	1 but NOT 9	β-receptors blocked	Thrower & Ord, 1974

Note: The markers affected (1–9) are shown in Table 7.1.

It is also evident that, for liver, the peak in cAMP at 12–13 h is not obligatory for transition to S phase. Indeed, in some hepatomas adenylate cyclase is absent. Otherwise the agents exhibit various ways of perturbing chromatin structure, either affecting the DNA (X rays, dimethylnitrosamine) or transcription from it (actinomycin, Be^{2+}, phenoxybenzamine). Excluding actinomycin, the effects of the blockers are selective; only chromatin associated with facultative gene expression is susceptible, not that from which constitutively expressed genes are being transcribed. We do not know what is the basis of this difference in sensitivity of the chromatin, but it is very striking that a protein kinase is inhibited by Be^{2+}, that X-irradiation selectively reduces histone H1 phosphorylation, and that, besides diminishing the induction of ornithine decarboxylase, an inducible nuclear kinase, utilizing histone H1 as substrate, is delayed in appearance by phenoxybenzamine.

The phosphorylations are catalyzed by protein kinases of different specificities (see Table 7.3), but all of them phosphorylate hydroxyamino acids with adjacent charged residues. The tyrosine protein kinase stimulated in oncogenesis and in normal cells by EGF, PDGF, and insulin is plasma membrane bound; otherwise the enzymes can all be located in the nucleus (as well as elsewhere) and can phosphorylate nuclear proteins. These nuclear substrates are potentially involved in charge interactions with other proteins or DNA, which would be perturbed by the introduction of new charged groups. It is possible that additivity/synergism between various growth factors could arise, at least in part, because of the complementary abilities of their dependent protein kinases to modify positively and negatively charged sites on chromosomal proteins associated with facultatively expressed genes.

TABLE 7.3. PROTEIN KINASES WHICH MAY BE INVOLVED IN PHOSPHORYLATIONS AT THE G_1/S RESTRICTION POINT

Kinase	Activator	Substrate Specificity	Reference
cAMP dependent	cAMP	—Lys/Arg—X—*Ser*—	Weller, 1979
Casein (nuclear) I	?	—Glu—X—*Ser*—	Section 5.2.4
Casein (nuclear) II	Polyamines	—*Ser/Thr*—Glu—	Section 5.2.4
Histone H1	?	—*Ser*—X—Lys	Romhányi et al., 1982
Tyrosine	EGF, PDGF, insulin	Glu/Asp—*Tyr*—	Baldwin et al, 1983

One further point is the probable necessity for tyrosine protein kinase activation for cell proliferation (Section 4.2.2). The functional importance and the substrates for this phosphorylation are unknown at the time of writing, and while specific inhibitors for tyrosine phosphorylation are being developed, results of their actions in normal cell growth have not yet been reported.

8

THE RESPONSES— THE NUCLEUS

For a cell in the G_0 state, differentiated and "doing its own thing" or quiescent, like small B lymphocytes, the stimulus consequent on tissue depletion provokes a major change in life-style. Additionally, in invertebrates and Amphibia, not only has replication and division to be set in train but, for many of the cells, the program to be pursued if regeneration is to be achieved is markedly different in detail from that of the original cells which contributed to the blastema. In some cells, such as enucleated erythrocytes, a change in cell program is not possible. Stem cells do not exhibit the highly differentiated properties expressed in their progeny, though it is thought that the potential to do so is already determined (Section 3.2.1). We will be concerned in this chapter with the way information stored in the nucleus is organized so that genes for proteins required continuously or periodically by the cell are available for transcription. We will also discuss the strategies used to enable highly specialized adult cells to pass on their differentiated properties to their daughters in environments far removed from those in the embryo, where the potential for those properties was first determined.

8.1. CHROMATIN ORGANIZATION AND ITS RELATION TO GENE EXPRESSION

8.1.1. Nucleosomes (McGhee & Felsenfeld, 1980; Igo-Kemenes et al., 1982)

The basic nucleosome structure is now known and is apparently universal. The 146 base pairs (bp) of DNA are coiled (1.75 turns) around an octameric core ($M_r = 11–15$ kDa), consisting of a tetrad of two molecules each of histones H3 and H4 together with two molecules each of histones H2A and H2B. One molecule of histone H1 (M_r c. 21 kDa) interacts with this structure at the site where the DNA duplex enters and leaves the core; when histone H1 is thus present, a further 20 bp of DNA become engaged, amounting in all to two complete supercoils. In histone H1, the central globular region has flexible N (40 amino acid residues) and C (c. 100 residues) terminal tails, the latter being especially basic. Globular regions of the core histones can be reconstituted to form a cluster after their basic N terminal sequences have been removed. It is uncertain the extent to which their basic N terminal arms are interlocked with DNA. Nuclear magnetic resonance (NMR) studies and the use of short cross-linking agents indicate that parts of histone H1 are close to histones H3 and H2A.

Isolated chromatin forms a relaxed linear assembly (diameter 10 nm), on which at low ionic strengths (< 60 mM NaCl) the nucleosomes are ordered, provided histone H1 is still bound. When the ionic strength is increased, the chromatin probably folds into solenoid forms, 20–30 nm across, containing six to seven nucleosomes (Thoma et al., 1979), possibly arranged edgewise to the DNA (McGhee et al., 1980) (for full discussion see Cartwright et al, 1982).

NUCLEOSOME SPACING

Almost all (see Section 8.1.3) nuclear DNA is organized into nucleosomes at all stages of the cell cycle. The nucleosomes are regularly spaced, as shown by electron microscopy and by nucleolytic cleavage with endonucleases, which preferentially attack the spacer DNA between the core particles. The size of the released fragments of DNA is subsequently determined by gel electrophoresis after ribonuclease treatment and deproteinization. The average repeat distance between nucleosomes varies (Thomas & Thompson, 1977) (chick liver with 200 bp, RBC with 212 bp), usually decreasing with increasing transcriptional activity. It also changes at different developmental stages (*Lytichinus* sperm with 248 bp, 64 cell stage with 213 bp; cerebellar neurons with 165 bp 4 days after birth to 218 bp by day 30; Jaeger & Kuenzle, 1982), with age (Zongza & Mathias, 1979), and possibly within the same cell (5 S rRNA genes in *Xenopus* with 175 bp, bulk chromatin with 189 bp) (see Igo-Kemenes et al., 1982). The mechanism by which spacing is determined is unknown; it may be closely associated with phasing. The significance of larger spacing is also unknown. Histones reduce the transcriptional potential of DNA and its susceptibility to nucleases. If transcriptionally less-active chromatin is mainly present in higher-ordered structures, the spacer DNA will be shielded (see McGhee et al., 1980) and the relative proportion of histone required to protect the DNA, reduced. Classical estimations of histone:DNA ratios by weight normally gave values of 1–1.5:1. Uncertainties inherent in these determinations would not give much significance to variations in the figures obtained.

Nonhistone chromosomal proteins (NHCP) (see Ord & Stocken, 1980; Bradbury et al., 1981; Cartwright et al., 1982) are also associated with nucleosomes and chromatin. The importance of the best-known set—the highly mobile group (HMG) of easily extractable, acid-soluble proteins (HMG 1, 2, 14, & 17; Goodwin et al., 1978) will be discussed shortly.

NUCLEOSOME PHASING (SEE KORNBERG, 1981; IGO-KEMENES ET AL., 1982)

Initially it was believed that nucleosomes were regularly spaced along the DNA, unrelated to its sequence or function. This is being intensively reexamined. The DNA studied has been principally reiterated genes such as those for 5 S rRNA (Gottesfeld & Bloomer, 1980), tRNA (Wittig & Wittig, 1979, 1982; Bryan et al., 1981), some satellite DNA (Igo-Kemenes et al., 1980), and SV40 (Huvasa et al., 1981), which is available in large amounts and has an easily recognizable replicatory origin closely adjacent to a unique Bgl I restriction site.

The experiments normally require nonspecific endonucleolytic cleavage of the chromatin to give suitable lengths of DNA. DNAase I, DNAase II, or micrococcal nuclease are commonly used; they are thought to be fairly, but by no means completely, nonspecific (Dingwall et al., 1981; McGhee & Felsenfeld, 1983). Sequence-specific cleavage is achieved at a limited number of sites by a suitable restriction enzyme and the resulting fragments end-labelled and examined electrophoretically. If the nucleosomes are phased, defined bands are visible; random arrangements lead to smears over the gels.

Evidence for phasing is probably least controversial for the SV40 chromosome (see Kornberg, 1981), for which it comes from a number of approaches. Here the replicatory origin is close to a unique Bgl I restriction site. This region is clearly visible in the electron microscope in about 20% of the SV40 particles, which show a gap in the regular nucleosome spacing (Saragosti et al., 1980). It is believed both that the nucleosome spacing is ordered from the replicatory origin and that the origin is bare of nucleosomes, at least when operational. With the *Xenopus* tRNA genes (Bryan et al., 1981) no indications of phasing were found in liver and cultured kidney cells, but in adult switched-off erythrocytes, phasing was detected and was linked to the high degree of chromatin condensation in the cells. Conversely, with chick embryonic tRNA genes which were being expressed, phasing was again detected (Wittig & Wittig, 1982) and ascribed to the need for the split sequences of the promoter site (5' to the start of the structural genes, and within the gene) to be accessible for transcription by RNA pol III.

If nucleosome phasing is confirmed, either in respect of replication and/ or transcription or in association with chromatin packing, it carries with it the implication of site-specific interactions between DNA and protein, to set the phase. Usually such a protein would be envisaged as nonhistone, but conceivably a minor histone variant (Section 8.4.2) is not precluded. The variability of nucleosome repeat distances implies some spacing mechanism, which is rather easily postulated as a component in the machinery of replication at the origins. If current suggestions (Jackson et al., 1981) are validated that replication and transcription occur from fixed sites associated with the nuclear matrix (see Nuclear Matrix, following), with moving DNA rather than moving machinery, parallels with engineering models for spatially separated packages are obvious. Changes in spacing and the acquisition or loss of phasing would require cells to pass through an S phase.

8.1.2.   The Nuclear Matrix

If metaphase chromosomes are isolated from HeLa cells and treated with dextran sulfate and heparin, all the histones and much of the nonchromosomal proteins can be removed to give highly folded loops of DNA (Marsden & Laemmli, 1979). This observation is the basis of the radial loop model of chromosome structure (Paulson & Laemmli, 1977). Here loops of DNA which might correspond to different domains along a chromosome are held in place by attachment to a central, nonhistone protein core. The composition and properties of the structure depend on the precise method of isolation used. About 30 different proteins are visible by sodium dodecylsulfate (SDS) polyacrylamide gel electrophoresis, with three predominating (Lebkowski & Laemmli, 1982). In the nuclear matrix from lymphocytes, lamin and actin have been identified (Nakayasu & Ueda, 1983).

In the radial loop model, a small, specific region of the DNA interacts with the anchoring proteins. Experimentally, most of the DNA can be removed by endonucleases, but the amount and type of nuclease resistant DNA depends on the enzyme used. With HeLa metaphase chromosomes and Hae III digestion, about 1.5% of the total DNA resists treatment, and restriction mapping shows this to include repeated sequences derived from human satellites II and III. If micrococcal nuclease digestion is used, only 0.1% DNA remains, containing no reiterated satellite sequences (Bowen, 1981).

Cells in interphase have yielded superficially similar structures. Lysis by nonionic detergents, followed by protein extraction with 2 M NaCl, gave "nucleoids" with loops of DNA which were still supercoiled. Partial digestion with restriction endonucleases left resistant DNA associated with small amounts of protein in cagelike structures (Cook & Brazell, 1975). If 8 M urea and 10 mM EDTA are used, more protein can be removed.

These structural observations have attracted considerable attention because of parallel experiments indicating that the nuclear matrix is a site where DNA synthesis is initiated (Long et al., 1979; Pardoll et al., 1980; Vogelstein et al., 1980). It has been associated, too, with tissue-specific transcriptional activity (Robinson et al., 1982). The findings are reminiscent of those in the 1960s which showed that DNA synthesis was initiated from specific regions in nuclei, then thought to have been linked to the nuclear membrane. Related experiments further demonstrated that some regions of heterochromatin replicated late (see Ord & Stocken, 1972b).

With the availability of hybridization techniques, it is now possible to show that the genes isolated with the nuclear matrix are those which are being selectively expressed in a particular tissue; for example, ovalbumin genes are preferentially bound to the matrix in chick oviduct cells but are not so bound in liver. Conversely, the globin gene is not associated with the matrix in oviduct cells and is not expressed (Robinson et al., 1982). In cell lines transformed with polyoma or avian sarcoma virus, integrated viral sequences which were being expressed in the transformed cell were also found to map with the nuclear cage (Cook et al., 1982) close to the points of attachment. Protein uniquely found in heterogeneous ribonucleoprotein (HnRNP) particles (see Section 8.1.3), which are part of the machinery for processing pre-mRNA, have been detected immunochemically on the nuclear matrix (Vogelstein & Hunt, 1982), as also have steroid hormone receptors, which are believed to bind to regions on DNA adjacent to the genes whose transcription is promoted (Barrack & Coffey, 1982).

If transcriptionally active DNA is specifically linked to the nuclear matrix, how is the correct selection of the genes achieved? Nonhistone proteins HMG-14 and 17 are specifically associated with transcriptionally active chromatin (see Section 8.1.3). The proteins are glycosylated, and Reeves and Chang (1983) found them to be present in the nuclear matrix; they were not bound if the glycosylated residues were removed. If these observations are confirmed, DNA chosen for binding by the matrix would be predetermined by chromatin components already selected because of their role in transcription. Further, the contiguity of the binding sites with those for the transcriptional machinery raises anew the possibility that signals transduced from the plasma membrane, for example, Ca^{2+} and cAMP, could be channeled via the cytoskeleton and microfilaments surrounding the nucleus to the nuclear matrix, and so regulate the machinery directly (Sections 6.1.2 & 6.4.1).

8.1.3   Transcriptionally Active Chromatin (see Mathis et al., 1980)

Techniques by which transcriptionally active chromatin could be selectively released from nuclei were devised empirically in the late 1960s. Many of these used sonication, in a variety of conditions which were not always understood, and actinomycin-D-sensitive pulse labelling with ^{3}H-uridine to detect newly formed RNA. Reproducibility and the type of RNA were therefore problematic. In the last few years hybridization with cDNA has become available for precise identification of the transcripts, and brief exposures (1–5 min) to the endonucleases DNAase I or II or micrococcal nuclease have become the usual means by which the actively transcribed chromatin is released. The nucleases preferentially attack double-stranded DNA between the nucleosome cores. DNAase II releases fragments of actively transcribed chromatin which are soluble in the presence of Mg^{2+} (Gottesfeld et al., 1974); this preferential solubility may require the chromatin to be hyperacetylated (see Sensitivity to DNAase I, following) (Goldsmith, 1981). With all three nucleases the release is tissue specific, preferentially occurring only in cells when the genes are being actively expressed (DNAase II, Gottesfeld & Partington, 1977; micrococcal nuclease, Bloom & Anderson, 1978; DNAase I, Wu et al., 1979).

Characterization of the properties of active chromatin has gone furthest for globin genes released by DNAase I (Weisbrod, 1982). If chromatin was first treated with 0.35 M NaCl, a number of nonhistone proteins were removed, and preferential sensitivity to DNAase I was lost but could be specifically restored by nonhistone proteins HMG-14 and 17 (Weisbrod et al., 1980). HMG-14 and 17 seem to be interchangeable; their binding is not prevented by the presence of histone H1 and appears to require the N terminal regions of the core histones (Mardian et al., 1980; Weisbrod, 1982). HMG-17 has been located in the linker region immediately adjacent to the nucleosome core (Sasi et al., 1982), where two molecules of HMG-17 are reported to bind between the entry and exit points to the nucleosome (Yau et al., 1983). The presence of HMG-17 significantly reduces the premelt region in thermal denaturation curves obtained with mono- to trinucleosomes (Yau et al., 1983).

If chromatin from chick leukemic cells was stripped by 0.65 M NaCl to remove the nonhistone proteins and histone H1, and then passed through affinity columns to which HMG-14 and 17 were bound, 5% of the chromatin was selectively retained, showing that the binding was a property still inherent in the stripped chromatin (Weisbrod & Weintraub, 1981). Further examination showed that (1) its DNA was undermethylated, (2) the extent of acetylation of its core histones was above average, (3) its histone H3 molecules dimerized more easily than in the bulk nucleosomes, and (4) topoisomerase I was a nonstochiometric component (Weisbrod, 1982).

HMG-14 and 17 have highly conserved sequences which are neither species nor tissue specific. Chromatins containing genes which are transcribed at different rates show the same affinity for HMG-14 and 17, and selective binding to transcriptionally active nucleosomes in vitro is achieved only if excess HMG-14 and 17 are present (Weisbrod, 1982). Amounts of HMG-14 and 17 in nuclei correspond rather exactly with amounts (10–20%) of transcriptionally active chromatin normally present. Notwithstanding these correlations, it is currently accepted that the criteria so far mentioned (and some others considered in the following) are not restricted to transcriptionally active chromatin. Taken together, however, regions with these properties are likely to be those engaged in transcription. Further features defining active chromatin may be expected (see Kuehl et al., 1980; Gurdon et al., 1982; Weisbrod, 1982; Nicolas et al., 1983.

Sensitivity to DNAase I

Illustrative of our present limitations in understanding is the selectivity of the effects of DNAase I on chromatin, a selectivity which is not seen with naked DNA. Two classes of site are now distinguishable (see Elgin, 1981): very sensitive (hypersensitive) sites, immediately 5′ to regions of transcription; and less-sensitive sites extending for several kbp both upstream and downstream from the 3′ end of the α- (Weintraub et al., 1981) and the β-globin gene structures (Stadler et al., 1980; McGhee et al., 1981). It is the hypersensitive sites which are correlated with, and can be reintroduced by, the presence of HMG-14 and 17 in the chromatin. The two patterns of nuclease sensitivity are being described in increasing numbers of cells, including the genes for *Drosophila* heat-shock proteins (Zachau & Igo-Kemenes, 1981), glue protein (McGinnis et al., 1983), histones, and chick conalbumin. In some viruses (Herbomel et al., 1981) and in the chick β-globin gene 5′ (McGhee et al., 1981), G-C–rich sequences are located in the hypersensitive regions, but they do not appear to be universally present.

Acquisition of hypersensitivity to DNAase I is being examined by Weintraub and his colleagues. In chick embryo fibroblasts, hemoglobin production can be induced by Rous sarcoma virus (rsv) and avian erythroblastosis virus. With rsv temperature-sensitive mutants it was demonstrated that DNAase I hypersensitivity was induced in parallel with the acquisition of hemoglobin expression. If the cells were transformed and then transferred to and maintained at the nonpermissive temperature for 20 generations and reexamined, the globin gene was not expressed, but DNAase I–hypersensitive sites were still evident. No change in DNA methylation was detected, though undermethylation of globin sequences was found if the cells were transformed by avian erythroblastosis virus (Groudine & Weintraub, 1980, 1982). The hypersensitive sites were susceptible to S1 nuclease, and Larsen and Weintraub (1982) therefore postulated they might be regions with single-stranded DNA. These could be self-replicatory if normally they were stabilized by protein preferentially binding to such DNA and available in excess.

The determinants for the lower, but still above average, DNAase I sensitivity of the more extended regions of the chromatin are also being studied, with the chick ovalbumin system (Stumph et al., 1983). Differential transcription from this domain occurs in the oviduct after exposure to steroid hormones, that is, it offers a model for inducible protein synthesis. Expression is associated with the acquisition of increased sensitivity (not hypersensitivity) to DNAase I over a region of about 100 kbp. Adjacent to, or just into, this region are certain middle repetitive sequences (Stumph et al., 1983). The significance of such sequences is provocatively speculative (Section 8.2.3).

Better understanding of higher-ordered chromatin structure, and the strategies by which gene domains are located in or at the periphery of such regions, should aid the interpretation of median DNAase I sensitivity. Hypersensitivity may entail greater knowledge of the basis and extent of nucleosome phasing and spacing and of the signals contained in DNA sequences for initiation by RNA pol II (Section 8.2.1).

DNA Methylation (Razin & Friedman, 1981)

DNA methylases in eukaryotes preferentially methylate cytosine situated $5'$ to guanosine—^mCpG. CpG sequences are deficient in vertebrate DNA (Josse et al., 1961), probably because 5^mC is a mutational hot spot. Not all CpG sequences are methylated. Methylated sites on DNA can be recognized by using isoschizomeric pairs of restriction enzymes, both of which preferentially hydrolyze a sequence, say CCGG, but only one of which attacks C^mCGG. Msp I and Hpa II are such a pair of enzymes, Msp I still being effective if cytosine adjacent to guanosine is methylated. Vertebrate DNA is much more heavily methylated than that in invertebrates, but at the same time it contains domains which are undermethylated. It is these latter which are associated with transcriptional activity (see Weisbrod, 1982; Cooper et al., 1983). In a number of cases, with the transcription of both pre-rRNA and pre-mRNA (see Cooper et al., 1983), undermethylation is evident at the 5' end of the gene but not through the whole of the sequence. With the ovalbumin and conalbumin genes and with the globin system, tissue-specific expression was correlated with undermethylation. In cells where the sequence was not expressed, DNA methylation was found.

As increasing numbers of sequences are examined in tissues at different developmental stages or with varying extents of specified gene transcription, it is becoming evident that correlations between hypomethylation and gene expression are not simple (e.g., Gerber-Huber et al., 1983; Vedel et al., 1983). Undermethylation around the 5' end of the structural gene may be a necessary but not sufficient condition for gene expression in higher eukaryotes.

One of the attractions of DNA methylation as a regulatory device is the ease with which, once established, it can be self-perpetuating (Riggs, 1975; Holliday & Pugh, 1975). Eukaryotic enzymes preferentially methylate sequences in which one strand of the duplex is already methylated (Bird, 1978), that is, duplexes are either not methylated:

$$-\text{CCGG}$$

$$-\text{GGCC}$$

or methylated on both strands:

$$-\text{C}^{m}\text{CGG}$$

$$-\text{GG}^{m}\text{CC}$$

Eukaryotic determinants for methylation *de novo* are awaited.

HISTONE ACETYLATION

Histone acetylation was first observed in the 1960s by Allfrey, Mirsky, and their colleagues and was soon linked with active transcription (for early references see Elgin & Weintraub, 1975). Histones H1, H2A, and H4 are stably acetylated on their N terminal serines very soon after synthesis; the core histones show additional variable acetylation on ϵ-NH_2 lysine residues. Histone deacetylation is inhibited by sodium butyrate (5 mM) (Candido et al., 1978), which causes a wide range of usually reversible cellular changes, including excessive histone acetylation.

There is complete agreement that the endonuclease attack on chromatin preferentially solubilizes regions containing highly acetylated histones H3 and H4 and acetylated H2B (Simpson, 1978; Perry & Chalkley, 1981). Fractionation of the released material after digestion is markedly affected by the Mg^{2+} concentration and extent of hyperacetylation (Georgieva et al., 1982). Many attempts have been made to demonstrate effects of histone acetylation on transcription in vitro. The results have been extremely conflicting, perhaps because of technical difficulties in using and maintaining transcription with endogenous RNA polymerase II. Current interest is based rather on the detection of actively transcribed gene sequences in the nuclease released material.

It was originally supposed the acetyl groups would significantly diminish electrostatic interactions between the positively charged lysine residues and DNA. We have already mentioned, however, uncertainties regarding the impediment which the N terminal tails that carry the acetyl residues (see Elgin & Weintraub, 1975) offer to transcription. In contrast to the stability of the methyl groups on DNA, ϵ-NH-acetyl groups on core histones turn over rapidly. Active turnover implies a cycle of reactions which are often regulated. Histone acetylation had long been known to vary through the cell cycle (for recent work using *Physarum* see Waterborg & Matthews, 1983). Histone acetylase, however, has not been shown to change in its activity through the cell cycle, and although histone deacetylase activity in *Physarum* rises sharply late in G_2 phase (Waterborg & Matthews, 1982), no regulation has so far been established. The changes seen in cell behavior when deactylation is inhibited demonstrate clearly, however, that the balance in enzyme activity must normally be controlled.

NUCLEAR PROTEIN PHOSPHORYLATION

Nuclear protein phosphorylation was detected about the same time as histone acetylation, and its investigation has followed a similar course (for earlier ref. see Elgin & Weintraub, 1975). From the start, histone phosphorylation was correlated with events in the cell cycle (Ord & Stocken, 1967). In cells in the G_0 state, however, phosphorylation, as monitored by $^{32}P_i$ uptake, is most evident on nonhistone proteins, where it has been especially examined in the group of proteins (M_r up to 35 kDa) extractable in 0.35 M NaCl, which includes the HMG proteins. Higher-molecular-mass proteins also incorporate ^{32}P$_i$ but are poorly characterized, only a few common components such as actin and tubulin being identified so far on the SDS gels. Only one of the histones, H2A, shows seryl phosphorylation in G$_0$ in actively transcribing cells (Marks et al., 1973); histone H2A has two phosphorylatable sites in its N terminal tail. If both of these are phosphorylated, 15–20% of histone H2A molecules in liver will be modified (Ord & Stocken, 1975b), as has been confirmed by more recent figures in a range of mouse and rat nuclei (Prentice et al., 1982).

The phosphate on histone H2A turns over very rapidly; the potential significance of this in transcription has been argued along lines similar to those discussed for acetylation (Ord et al., 1975; Allis & Gorovsky, 1981; Prentice et al., 1982). It is noteworthy that simple phosphorylation is seen only on histone H2A in the G$_0$ state, and this is the core histone which is least significantly acetylated. An association between histone H2A phosphorylation and transcriptionally active chromatin has been demonstrated, though less elegantly than for acetylation (Ord & Stocken, 1979, 1981b; Prentice et al., 1982).

At least five different protein kinases are found in nuclei (Sections 5.2.4 and 7.4), most of which can be regulated. Their range of specificity gives a potential for finely controlled changes. Apart from precise effects which phosphorylation can have on functional molecules such as phosphorylases, two different consequences of nuclear protein phosphorylation have been postulated—a weakening of DNA-protein interactions (cf. acetylation) and, for histones H1 and H2A, a need for phosphorylation in intermolecular interactions in mitosis or heterochromatinization (Section 8.1.4). Dephosphorylation is believed to be brought about by protein phosphatases currently being studied from the cytoplasm (see Cohen, 1980; Ingebritsen & Cohen, 1983). Neither inhibitors nor mutants are available at present to enhance or diminish nuclear protein phosphorylation that have comparable effects to those which butyrate has on acetylation. It is therefore much harder to provide conclusive rather than circumstantial evidence of in what respects this class of modification to nuclear proteins matters.

ADP RIBOSYLATION

ADP-ribose is derived from NAD; it is a modification introduced into glutamate residues (NH$_2$OH-sensitive), seryl phosphate (NH$_2$OH-insensitive), and possibly other amino acid residues by the enzyme poly(ADP-ribose) polymerase (see Hilz & Stone, 1976; Hayaishi & Ueda, 1977; Purnell et al., 1980; Ord & Stocken, 1980). ADP-ribose derivatives exist in two forms, monomers and highly flexible polymeric attachments (Hayashi et al., 1983). ADP-ribosylation occurs widely in nuclei, mitochondria, and cytoplasm (Adamietz et al., 1981); it is prominent in the plasma membrane in the presence of cholera toxin and on elongation factor-2 in the presence of diptheria toxin (see Vaughan & Moss, 1981). In nuclei both histone and nonhistone proteins can be acceptors for ADP-ribose. The involvement of the two classes of protein differs between mono- and polyADP-ribosylation. Immunochemical evidence suggests nonhistone proteins may be the more important for polyADP-ribosylation (Kun et al., 1983).

The function of the modifications is uncertain, but increasingly, nuclear polyADP-ribosylation is being associated with DNA repair (Durkacz et al., 1980). The proportion of NH_2OH-resistant monoADP-ribosylated proteins increases in terminally differentiated cells (Hilz et al., 1979; Bredehorst et al., 1981). In liver nuclei from adult cells in the G_0 state, 5–10% of histone H1 is monoADP-ribosylated (Ord & Stocken, 1975b). Mono ADP-ribosylated histone H1 is selectively released by micrococcal nuclease in association with transcriptionally active nucleosomes (Caplan et al., 1978; Smulson, 1979). There has been a substantial number of reports that nucleosomes released from transcriptionally active chromatin are depleted in histone H1 (Egan & Levy-Wilson, 1981; Gabrielli et al., 1981). The ease with which the histone is dissociated, especially if it is ADP-ribosylated, suggests that the loss may be artefactual.

Smulson and his colleagues (Butt et al., 1978; Giri et al., 1978), from their findings that poly(ADP-ribose) polymerase was preferentially located at linker regions in poly (8–10) nucleosomes, suggested ADP-ribosylation was involved in maintaining higher orders of chromatin structure. When purified poly(ADP-ribose) polymerase was added with NAD to polynucleosomes in vitro, polyADP-ribosylation of histone H1 occurred with relaxation in structure rather than aggregation (Poirier et al., 1982). Reproducing appropriate conditions for assembly in vitro is, however, notoriously difficult.

8.1.4. Nucleosome Organization and Higher Chromatin Structures

A full discussion of the structure of transcriptionally active chromatin, derived from many studies using electron microscopy and nuclease digestion, is to be found in an extensive review by Mathis and colleagues (1980).

Ultrastructural studies have shown that nucleosomes are usually present at sites where HnRNA transcription by RNA pol II is occurring. With nucleoli, however, especially in electron microscope (EM) studies through amphibian oocyte development, fern-leaf patterns of newly transcribed RNA are seen, with nucleosomes absent from the sites of active transcription by RNA pol I (Scheer, 1978; Trendelenburg & McKinnell, 1979). Inactive regions of the nucleoli, where transcription has not been recruited, show normal nucleosome organization.

Further indications for altered nucleosome organization around ribosomal RNA genes have come from micrococcal nuclease and DNAase I digestion patterns of yeast chromatin 35 S ribosomal RNA genes (Lohr, 1983). Nucleosome patterns changed abruptly around sites for the initiation of transcription and 5' flanking sequences upstream. The authors suggested that, where very active transcription by RNA pol I was occurring, normal nucleosome arrangement was absent, both from the upstream control sites and from the structural gene region. If normal nucleosomes were present in the structural gene region but not around the control sites, transcription was possible, and if both regions shared normal nucleosome patterns, transcription was switched off.

The structure of heterochromatin has hardly started to emerge. This form of chromatin was originally described from cytochemical studies, heterochromatin being the more densely staining nucleoprotein in contrast to euchromatin. Later autoradiographic studies showed euchromatin to be transcriptionally more active.

For some time the importance of histone H1 in building up higher-ordered structures has been agreed; cross-linking studies have now detected homopolymers of up to 12 molecules of the histone (see Igo-Kemenes et al., 1982). In the most condensed stage of chromatin, at metaphase, phosphorylation of histone H1 is thought to be essential for increased intermolecular interactions (see Matthews, 1980).

The hydrophobicity of the globular regions of histones H2A, H2B, H3, and H4 is involved in maintaining the association of the octameric core of the nucleosomes. Hydrophobic variants of histone H2A were reported to be relatively increased in heterochromatin in cell lines from the mouse *Peromyscus*, and this histone was substantially phosphorylated. Phosphatase activity in the preparations was difficult to control, and turnover data are not available for the phosphate groups compared to those on histone H2A in transcriptionally active chromatin (see the preceding subsection on Nuclear

Protein Phosphorylation). As expected, extents of acetylation of histones H3 and H4 in heterochromatin were reduced (Halleck & Gurley, 1982).

A related problem, about which we know almost nothing at present, concerns determinants for heterochromatinization, which is established early in development and replicated through daughter cells with great fidelity (Sparfford et al., 1978). The structural basis for heterochromatin associated with highly reiterated gene sequences around centromeres is under investigation; the nonhistone proteins present are difficult to analyze because of their extreme insolubility (see Bradbury et al., 1981). Translocation of genes to sites adjacent to heterochromatic regions causes them to be switched off. In *Drosophila* salivary glands, mosaicism can occur, with genes being active in some adjacent cells but not in others, indicating the importance of highly localized effects during determination.

8.2. THE TRANSCRIBED MESSAGE

8.2.1. Signals for Transcription

We have already seen that pre-mRNA processing appears to be a prominent site of transcriptional control in early stages of adult somatic cell growth (Section 7.2.1). There is no evidence that RNA polymerase availability is a critical limitation in any cell type in G_0 phase, though the synthesis of more molecules of the polymerases may be part of the coordinated response to growth-promoting signals (Section 7.2.2). RNA polymerase binding and initiation of transcription are therefore thought to be control stages, determining the transcriptional expression in the phenotype and potentially being regulated through the cell cycle.

Classical experiments in the 1960s showed that histones restricted transcription from DNA. Nevertheless, present views of chromatin structure with DNA coiled around the histone core suggest that some of the genome is accessible, at least in the relaxed, open form of chromatin (Weintraub, 1980). A peptide with 50 amino acid residues, M_r c. 6 kDa, could be encoded by a DNA sequence of 150 bp, that is, approximately the length of a DNA duplex associated with one core particle. Core histones themselves, whose genes are commonly without introns (Kedes, 1979; Hentschel & Birnstiel, 1981), contain 100–140 amino acid residues which could be encoded by the DNA associated with two nucleosomes. Globular proteins with M_r = 60–70 kDa require about 2 kbp coding information; with intron:exon ratios rapidly increasing as genes for bigger proteins are sequenced (2:1 upward toward 10:1), sequence availability and topological difficulties in organizing transcription and subsequent processing are becoming acute problems.

With the discovery in the 1970s of middle reiterated DNA sequences and their frequent interspersion between single-copy DNA, it was realized that various classes of regulatory or functional DNA sequences might show these hybridization kinetics. They would include binding sites for the three different RNA polymerases and, rather less abundantly, sites for steroid hormone receptor binding, possibly others for cAMP regulatory protein binding, and so on. These and other repeated sequences, homologous or closely related to those having functional significance in prokaryotes or viruses, are continuously being reported (see Jelinek & Schmid, 1982; Singer, 1982). We can only mention those so far identified of most direct importance to normal growth.

When RNA polymerase II is bound to DNA, about 40 bp are protected. Consensus sequences which are believed to be involved in transcriptional signals are shown in Table 8.1. The TATA box 25–30 bp upstream from the 5' end of structural genes is reminiscent of the run of sequences required for prokaryotic messengers, the Pribnow box. The importance of the A-T–rich sequence confirms earlier, less-direct evidence (see Flickinger, 1982). The early gene region in SV40 virus is notable for the absence of the TATA box but shows multiple, specific sites for initiation. As well as the TATA box, sequences further upstream are also implicated (see Breathnach & Chambon, 1981 & Table 8.1) with others for termination 3' to the structural gene. Additionally, quite short, common cis sequences are being found upstream from structural genes coding for unlinked but coordinately regulated proteins, including some of the genes for heat-shock proteins in *Drosophila* and for glucocorticoid controlled gene functions in human and rodent cells (Davidson et al., 1983).

Glucocorticoid regulation can be imposed on gene functions by the insertion of the appropriate DNA sequence upstream to the structural gene whose expression is to be controlled. Steroid hormone receptor complexes bind to long terminal repeat (ltr) segments of mouse mammary tumor virus. About 270 bp from the ltr become incorporated into the host genome and, on treatment with dexamethasone, viral transcription is enhanced (Lee et al., 1981; Govindom et al., 1982). Similar experiments have been performed

TABLE 8.1.   PROBABLE CONSENSUS SITES FOR EUKARYOTIC mRNA

Initiation (see Breathnach & Chambon, 1981; Paul, 1982; Dierks et al., 1983)

mRNA start site

$$5' \quad G\text{—}GTATA^A_TA^A_T\text{—}G\text{——}G \xleftarrow{\text{9–17 bp}} Py\text{-----}Py\,\overset{\downarrow}{A}\,PyPyPyPy$$

$$\xleftarrow{\qquad \text{Approx. 25 bp} \qquad}$$

Consensus Sequences

$$5' \qquad \text{—}GG^C_T CAATCT\text{—}$$

Approx. position: 80   70

$$5' \qquad PyPuPuPu\text{—}CC^A_C T CACCTG$$

Approx. position: 100

Termination (Proudfoot & Brownlee, 1976)

$$5' \qquad AATAAA \xleftrightarrow{\quad 14\text{–}20\ bp \quad} \text{—}G\text{—}\overset{\uparrow}{C}\text{—}poly\ (A)3'$$

Usual but not invariant

Junctional Sequences in Genes with mRNA Splicing (see Mount, 1982)

$$^C_A AG/GT^A_G AGT\leftarrow \text{Consensus donor (exon-intron) boundary}$$

$$^T_C N^C_T AG/G\leftarrow \text{Consensus acceptor (intron-exon) boundary}$$

with a thymidine kinase (tk) gene adjacent to the ltr: integration into tk⁻ host cells and treatment with dexamethasone enhanced the production of thymidine kinase (see Cato, 1983).

Another set of consensus sequences thought to be present in all mRNA carrying introns which have to be excised and spliced are those at the junctions where the incisions occur (Mount, 1982; see Table 8.1). Breathnach and his colleagues (1978) had earlier observed what is emerging as a rule for mRNA genes; introns begin with GT and end with AG. Splicing junctions appear to be highly conserved.

Signals to RNA pol I to bind to rRNA genes (Long & Dawid, 1980) are again thought to be 5' to the start of the structural sequence; there is no

TATA box, but there is a T-rich sequence—20-25 bp upstream (Moss & Birnstiel, 1979). RNA pol III (5 S RNA and tRNA) has a split site, including residues near the 5' end of the structural genes and others 50–60 residues downstream into the genome (Galli et al., 1981; Hofstetter et al., 1981).

Initiation of transcription requires protein factors additional to the polymerases (for RNA pol II, Davison et al., 1983; for RNA pol III see Marx, 1981). Purified RNA pol II can be phosphorylated by nuclear kinase II; initiation is then increased (Stetler & Rose, 1982). Whether this occurs in vivo is unknown. Complexes containing chromatin-bound RNA pol II have been isolated following endonuclease treatment (Sargan & Butterworth, 1982; Yukioka et al., 1982). These may aid in characterizing the proteins required in initiation and ultimately allow its reproducibility in vitro.

8.2.2. Ribonucleoprotein Particles (*HnRNP Particles*)

If chromatin is lightly digested with endonuclease, pulse labelling or hybridization studies show nascent mRNA is associated with the released chromatin. RNP particles carrying the newly synthesized RNA can be dissociated from the chromatin in isosmolar NaCl at pH 8.0 and isolated by sedimentation through sucrose gradients (Louis & Sekeris, 1976). The particles contain stable, small RNA (snRNA, 100–300 nucleotides) (Zieve, 1981; Zieve & Penman, 1981) and about six distinct proteins ($M_r = 32-44$ kDa) (Beyer et al., 1977; Webb-Walker et al., 1980). It is believed that pre-mRNA are processed in these HnRNP particles and that snRNA may be involved in orienting the pre-mRNA for splicing.

The proteins in the HnRNP particles may be phosphorylated and can be ADP-ribosylated in vitro (Kostka & Schweiger, 1982). Protein phosphorylation in HnRNP particles from liver was enhanced when dexamethasone was given to rats to promote enzyme induction and diminished when this process was inhibited (Ord & Stocken, 1981a). We have already reviewed the acceleration in pre-mRNA processing when growth is promoted (Section 7.2.1).

8.2.3. Strategies in Differentiation

Experiments with nuclear transplantation in the 1950s (see Section 1.1.1) have been interpreted as showing that, in the adult somatic cells used, differentiation had been achieved without irreversible changes in the genome.

At the time it was already known that some types of differentiation proceeded contrary to this, with losses of DNA as in mammalian erythrocytes and the XO *Drosophila*. In the last few years increasing numbers of cell types have been found to employ differentiational strategies where the DNA sequences are changed, either by gene amplification, gene deletion, and/or gene rearrangement.

Gene amplification is well established for r and t RNA genes (see Long & Dawid, 1980). It is at least partly reversible; at meiotic reduction in amphibian oocytes, highly amplified rRNA genes are reduced in number (see Gerhart, 1980). Gene amplification can be produced in eukaryotic cells by various experimental procedures (Dolnick et al., 1979). There is no evidence at present that amplification of selected genes in different tissues is a common differentiational device.

Elimination of DNA sequence is found during the maturation of B lymphocytes, associated with the expression of a different class of heavy chain without change in the variable region of the immunoglobulin. Genes for the constant region of the heavy chains are linearly arranged $-\mu-\delta-\gamma-$; as expression switches from the IgM class to that of IgG; for example, DNA sequences $5'$ to the gene are deleted (see Marcu & Cooper, 1982). Rearrangement of gene sequences involved in immunoglobin light chain production was first detected from the closer spacing of the genes for the variable and constant regions of, for example, a λL chain, in a cell where the light chain was being expressed when compared with a cell not so committed (see Eisen, 1980; Siebenlist et al., 1981; Leder, 1982).

The discovery that viral gene products, notably the tyrosine protein kinase, may be closely related to or identical with those of normal host genes has prompted speculation that exchange of genes between host and plasmid may have been a significant occurrence over evolutionary time. Following from this, the question arises of the extent to which translocation of genes, and especially their shift to or from a site influenced by promoters, may be a normal mode of differentiation. There is sufficient information about analogous prokaryotic transpositions to hypothesize about junctional sequence requirements for such translocations, but there is no knowledge of the conditions which would allow rearrangement of DNA sequences to occur. Reversibility, which is a necessary condition for tissue regeneration in the absence of stem cells, is hypothetical but improbable.

Because reversibility of differentiation by alteration of DNA sequence seems unlikely within the time span of the regeneration process, we will turn next to the different strategies so far postulated for differentiation without change in DNA sequence. The most obvious of these is the use of DNA sequence-specific repressor proteins, as in *E. coli* and the *lac* operon. The extremely small amounts of nonhistone protein which would be required for single or low-copy-number genes may be the reason for our failure, as yet, to identify these proteins. It is also possible that, in higher vertebrates, if classical repression is established during embryogenesis, the presence of the repressor protein serves as a site to which a DNA methylase which will methylate DNA *de novo* is preferentially bound. If methylation can be facilitated in this way, it may be self-perpetuating, as we have already seen (Section 8.1.3), and perhaps allow the persistence of a repressed state of the genome without the need for continued synthesis of the repressor molecule.

A second device by which part of the DNA sequence can be distinguished is its adoption of an alternative conformation. Z-DNA is a left-handed conformation of the DNA helix (cf. the more usual right-handed B-DNA) with purine bases rotated around the glycosidic bond into a syn conformation. It was detected crystallographically in the deoxyribonucleotide CpGp-CpGpCpG and can be assumed when there are runs of alternating Pu and Py (see Nordheim et al., 1981). The molecule is strongly immunogenic. When antibodies were isolated and a fluorescent marker attached, Z-DNA was detected in *Drosophila melanogaster* polytene chromosomes at the interband regions (Nordheim et al., 1981) and in nuclei from many, but not all, rat tissues (Morgenegg et al., 1983). It is postulated that the sequence would require stabilization to prevent its ready reversal to the B form. Proteins selectively stabilizing Z-DNA have now been detected in *Drosophila* nuclei (Nordheim et al., 1982). Methylating cytosine to give 5^mC also stabilizes the Z form. It will be more difficult to detect Z-DNA in non-polytenized chromosomes, but provisional identification of potential Z-DNA at ends of the yeast chromosome has been reported (Walmsley et al., 1983).

Primary transcription, additionally to restricting transcription from various sequences, can be varied by using alternative capping or polyadenylation sites to yield different types of pre-mRNA from the same gene domain, which may then be variably spliced (see Darnell, 1982; Zehner & Paterson, 1983). Alternative transcripts from the α-amylase gene domain (haploid genome, single-copy gene Amy-1A) are produced in mouse liver and salivary gland (Hagenbüchle et al., 1981; Tosi et al., 1981). Two possible promoter sites have been identified, the stronger being exclusively active in the salivary gland (Schibler et al., 1983). The two mRNAs have identical coding and 3' noncoding regions, with tissue-specific leader sequences. Tissue-specific differences between the thyroid and hypothalamus have been reported, too, for the calcitonin gene domain (Amara et al., 1982) and for α_{2u}-globulin expression between male rat liver and submaxillary gland. With these proteins, transcription in liver is under endocrine, age, and sex control but is constitutive in submaxillary gland (Laperche et al., 1983). A further trick is gene switching as shown with respect to, for example, human globin genes. Here transcription is switched from the fetal γ-sequence to the β-sequence further downstream in late uterine life (Fritsch et al., 1980).

We have already assumed that the potential restrictions in gene expression, by means such as those outlined earlier in this section, are imposed during embryogenesis. The signals that allow alternative transcription through a gene domain, and those promoting gene switching, are not understood at the time of writing. It is possible that they include signaling mechanisms we have already reviewed (Chapter 4)—alterations in phosphorylation status, redox state, Ca^{2+}, and so on. Further, a process which acts so that a gene domain is transcribed to give one gene product under one set of circumstances (e.g., calcitonin in the thyroid) and another if the conditions are different (calcitonin gene related products in the hypothalamus, Amara et al., 1982) may be a useful model for the cell transformations occuring in limb regeneration within the fibroblast lineage.

These discussions have posited transcriptional control because differentiation has to be transmissible to daughter cells (see Table 8.2). We know, however, that cytoplasmic factors can produce selection of the mRNA translated (see Section 7.1.3) and that posttranslational changes can occur to protein precursor molecules, for example, proopiomelanocortin (see Section 4.2.1), which are cell type specific leading to different end products in different cell lines (Acher, 1980; Herbert & Uhler, 1982). The same uncertainties about controls arise as for alternative transcriptional mechanisms.

In summary of this section and review of the effects of known strategies for differentiation on the potential for regeneration (see Table 8.2), where mechanisms of phenotypic selection have not entailed alterations in DNA sequence, the possibility exists of tissue reformation. The recognized mechanisms may also be those which are affected in facultative gene expression (modulation, Weiss, 1973). We have considered how intracellular environmental changes transiently elicited by the signals described in Chapter

TABLE 8.2. DIFFERENTIAL STRATEGIES AND THE POTENTIAL FOR REGENERATION (See Sections 8.2.3 and 1.2)

Strategy	Differentiation — Mechanism	Differentiation — Example	Regenerative Potential	Modulation
Irreversible changes in the genome	Nuclear elimination	Mammalian RBC	*Impossible*	
	Chromosome elimination	XO *Drosophila*		
	Partial elimination of DNA sequence	Immunoglobulin biosynthesis		
Changes in the genome	Sequence rearrangement / Sequence amplification	Immunoglobulin biosynthesis / Amphibian oocyte maturation		
	Classical Differentiation / Persistent Environmental Factors Allowing / Transmissible Phenotypic Selection			
Transcriptional selection	Gene selection ± processing / Gene processing	Fetal to adult hemoglobin switch / α-Amylase gene	*Possible*	*Modulation (Weiss, 1973) / Facultative Gene / Expression Transient / Environmental Changes*
Translational selection	Initiation mRNA (binding onto ribosome)	α- and β-globin genes in reticulocytes		
Translational processing	Proteolytic cleavage	Proopiomelanocortin		

4 may influence protein synthesis. It is not known whether the same intracellular changes persist continuously to maintain the differentiated state, for example, high cAMP levels, or whether the modulation is a prelude to further changes, for example, DNA methylation, which cause the changes in gene expression to be stabilized. In the latter event it may be surmised that dedifferentiation (Section 1.2) is unlikely; in the former, de- and redifferentiation within a cell lineage appear to remain possible.

8.3. CHROMATIN CHANGES AS CELLS LEAVE THE G_0 STATE

There has been little recent work in this important area, even since the realization that the prominent early changes in gene expression involve more efficient processing and transport of normal transcripts, rather than transcription from new regions of the genome (Section 7.2.1). Further, the characterization of active chromatin, and its release from nuclei by relatively defined techniques, imply that some of the classical observations made with whole nuclei or total chromatin require reappraisal, to check if the early changes are confined to preexisting domains of active chromatin. It is also important to look for changes in the components or functioning of the HnRNP particles, where changes in the pre-mRNA processing may be localized.

Indications of alterations in the state of chromatin can be detected within 30 min of stimulation in a number of cell types (see Ord & Stocken, 1980). Acridine orange binding is enhanced. Acridines intercalate between the stacked bases of the DNA helix; ordered chromatin binds only minimally to acridine orange. The increased binding was interpreted as a "loosening" of chromatin structure, commensurate with the potential for increased transcription. Two further indications of changes early in the G_0/G_1 shift are increased nuclear protein acetylation and phosphorylation, and, in some cells, the appearance in nuclei of increased amounts of nonhistone protein.

8.3.1. Nuclear Protein Modifications as Cells Leave the G_0 State

Acetylation of core histones and phosphorylation, especially $^{32}P_i$ uptake by nonhistone proteins, is increased within 1–2 h of stimulation in regenerating liver, lectin-promoted lymphocytes, and a wide range of cell cultures (for early references see Elgin & Weintraub, 1975; Ord & Stocken, 1980). Increased modification could be due to an alteration in the balance of activities of modifying to demodifying enzymes or to altered accessibility of the protein substrates in the nucleus. It is known that the latter can be very important and can be influenced by the ingress of nonhistone proteins and by Na^+ (see Fonagy et al., 1977). It remains uncertain if variations in the Na^+ flux across the plasma membrane (Section 5.2.2) get reflected in Na^+ concentrations in nuclei.

Most attention has focused on alterations in the balance of enzyme activities. The detection of peptide inhibitors for deacetylation (Reeves & Candido, 1979) suggested that the hydrolase might be regulated, but this has not been found in cells in interphase (Waterborg & Matthews, 1982). There is at present no indication of whether acetylation or deacetylation are themselves modified in response to the secondary messengers already discussed.

In transcriptionally active chromatin, histone H4 may carry up to four acetyl groups. The acetylated forms are elegantly separable on extended gel electrophoresis (see Davie & Candido, 1978). There is little evidence that amounts of chromatin susceptible to nucleases increase through G_1 phase (e.g., Caplan et al., 1978). Precise quantitative comparisons of amounts of the acetylated species of histone H4 in active chromatin are difficult, and as we noted above, the significance of the correlation between increased histone acetylation and transcriptional activity is still unclear.

Phosphorylation of histones can also be analyzed electrophoretically, as well as by exclusion gel and ion exchange chromatography (Ord & Stocken, 1975b; see Fig. 8.1). Phosphate groups/mole protein and the proportion of phosphorylated:nonphosphorylated molecules can therefore be separately determined. Histone H2A carries only two phosphate groups when it is maximally phosphorylated in vivo (serine-1 & 19); molecules carrying two phosphate groups are detectable in chromatin from resting liver and lymphocytes (Ord & Stocken, 1975b; Fonagy et al., 1977). By 4 h after partial hepatectomy, the average extent of phosphorylation on histone H2A has risen from about 0.25 mole P/mole protein to 0.4. No indication was found that the proportion of molecules carrying phosphate increased early in G_1 phase, so the rise in phosphate must have been due to an increased number of molecules which had been carrying one phosphate and now carried two.

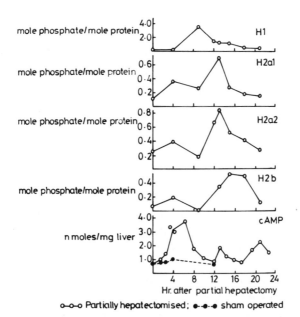

Figure 8.1. Phosphate content of rat liver histones following partial hepatectomy (Ord et al.). Reprinted by permission of the Plenum Publishing Corporation (*Subcellular Biochemistry*, Vol. 4, p. 150, 1975). (Histones H2a1 and 2a2 now redesignated H4 and H2A).

Further information is available from analyses in liver. Turnover of the phosphate on histone H2A is very rapid. After 1 h labelling in vivo in normal rats, the specific radioactivity of the (average) alkali labile, serine-bound phosphate was 60–70% of that in liver intracellular P_i (Ord & Stocken, 1975b). Four hours after partial hepatectomy the relative specific activity of the phosphate on histone H2A was approximately halved, showing that the ratio of kinase:phosphatase activity had changed. No detectable alteration in nuclear protein phosphatase activity has been reported up to early S phase (Ord et al., 1975). Levels of cAMP reach a peak 4 h after partial hepatectomy (see Fig. 5.1), so cAMP-dependent protein kinases are activated. These enzymes have been implicated in the phosphorylation of nonhistone proteins, of (monoADP-ribosylated) histone H1, and histone H2B, which also have increased phosphate at this time (Ord & Stocken, 1975b). Histone H2A is not a good substrate for cAMP-dependent protein kinases in vitro. Be^{2+}, which inhibits nuclear kinase I (Kaser et al., 1980), diminishes the phosphorylation of histone H2A in liver by about 33% both in G_0 state and 4 h after partial hepatectomy, thus implicating nuclear kinase I in the phosphorylation.

Increased phosphorylation of histone H2A in G_1 phase was also promoted in synchronized L cells by dexamethasone, phosphorylation of histone H1 showing no change (Prentice et al., 1978).

8.3.2. Changes in Nuclear Nonhistone Chromosomal Proteins

Transcriptionally inactive cells may have relatively small amounts of nuclear nonhistone proteins (Section 1.1.2). In small lymphocytes the ratio of DNA:histones:nonhistone proteins is 1.0:1.0:0.4, compared to 1.0:1.4:1.2 for liver (this laboratory's unpublished data). Increased nonhistone protein was found in nuclei one hour after W138 cells had been transferred into fresh serum (Baserga, 1976). The nonhistone proteins were detected by changes in circular dichroism spectra, showing the proteins had penetrated close enough to the DNA to affect its ellipticity. One of the unresolved problems is how these nonhistone proteins, which are present in the cytoplasm in G_0 cells, are selectively retained in nuclei after growth has been promoted. Microinjection studies into oocytes (Bonner, 1975) have examined the selectivity of nuclear retention; small basic proteins like lysozyme or cytochrome c were not retained, whereas similarly sized, basic histones were (de Robertis, 1983).

Further experiments with oocytes and the distribution of proteins between nucleus and cytoplasm also indicated selectivity in binding by the nucleus (Feldherr & Ogburn, 1980). Similar conclusions were reached for the distribution of nucleoplasmin, a protein which is thought to be involved in nucleosome assembly in *Xenopus* eggs. Monomers of the protein have relatively small core and a tail (M_r c. 12 kDa) which can be deleted enzymically. The tails penetrated into the nucleus freely, whereas the core did not (Dingwall et al., 1982). Boogaard and Dixon (1983) introduced mRNAs for protamines into HeLa cells by fusion. The mRNAs were translated and protamines II and III identified. Protamine II, but not III, was phosphorylated and localized in the nucleus of the HeLa cells but was less effectively bound that in normal, homologous, trout testicular nuclei. Once more, therefore, nuclei showed selective retention of one out of a pair of very similar molecules. The authors noted that phosphorylation might have been a factor in the retention—a suggestion that has also been made from studies of protein uptake by thymus nuclei (Ord & Stocken, 1980).

Besides the rapid entry of existing nonhistone proteins into nuclei as cells leave their G$_0$ state, new nonhistone chromosomal proteins are synthesized in G$_1$ in response to growth stimulation and migrate into the nuclei (Levy et al., 1973; Pogo & Katz, 1974).

The proteins which enter nuclei as cells move out of their G$_0$ state have not been precisely identified, but electrophoretic analysis has suggested that not many of the nonhistone proteins are cell type specific. If retention of some nonhistone proteins in nuclei is energy dependent (see discussion in Dingwall et al., 1982), the complement of the proteins may be reduced in nonproliferating cells as an energy-conserving adaptation.

8.4. EXTRACELLULAR EFFECTS ON CHROMATIN: G$_1$/S AND LATER

In Chapter 7, cytoplasmic changes in response to extracellular signals were considered. The population of cells with which we are concerned, once they have passed through the G$_1$/S restriction point (Section 7.4), commonly proceed to replicate and divide without interruption. A restriction point can be identified late in G$_2$ phase, but normal adult somatic cells do not arrest at this point. Indeed, with cells in culture, serum dependence is observed through G$_1$ but not thereafter (Stiles et al., 1981), and it seems, therefore, that nuclear changes as cells pass through the G$_1$/S restriction point may be affected by secondary messengers. Later stages of the cell cycle will be independent.

8.4.1. Nuclear Protein Modifications

PHOSPHORYLATION

Enhanced phosphorylation of histone H1 as DNA replication commenced (see Fig. 8.1) was first observed in regenerating rat liver (Ord & Stocken, 1967) and has so far been found in all cell types examined. Phosphorylation occurs onto serines in the carboxyl portion of the molecule (Hohmann et al., 1976); it usually slightly precedes [3]HTdR uptake in DNA and occurs first on "old" histone H1 molecules. Ion exchange analyses have shown that newly synthesized histones H1 then became more phosphorylated than in the bulk chromatin (see Ord & Stocken, 1975b). The [32]P$_i$ uptake and phosphate contents are also elevated for the other histones, except histone

H3, as DNA synthesis commences (see Fig. 8.1). Phosphorylation of histone H1 variants ($H1^0$, d'Anna et al., 1980a,b) accompanies that for H1. Phosphorylation of nonhistone proteins is selective (Ballal et al., 1975); HMG-1 and 2 do not show cell-cycle-dependent phosphorylation (d'Anna et al., 1980a,b; Bhorjee, 1981), but this has been reported for HMG-14, which showed a sevenfold increase in G_2 (Bhorjee, 1981; Bhorjee et al., 1983; Walton & Gill, 1983).

The basis of the enhanced phosphorylation is not entirely clear. Although a small peak in cAMP is found when phosphorylation of histone H1 first increases 12–15 h after partial hepatectomy (compare Fig. 5.1 & 8.1), the universality of the phosphorylation response, even in cells without adenylate cyclase, suggest cAMP-dependent protein kinases are not the only enzymes involved in the response. Polyamine activation of nuclear (casein) kinase II has already been considered (Section 7.4), as has the potential involvement of nuclear (casein) kinase I. Increases in the activity of chromatin-bound, cAMP-independent protein kinases have also been described (Siebert et al., 1971; Sons et al., 1976; Laks & Jungmann, 1980), possibly associated with new synthesis of the kinase early in S phase (Thrower et al., 1973).

Dephosphorylation, at least of the histones, is evident by the end of S phase (see Fig. 8.1); it apparently accompanies maturation of nucleosomes (Ord & Stocken, 1979). Extensive phosphorylation of histone H1 occurs at the end of G_2, as a prelude to mitosis (Gurley et al., 1973), probably through activation of preexisting, growth-associated histone kinase (Matthews, 1980). Histone H3 is selectively phosphorylated during mitosis; again the basis of this is unknown.

ACETYLATION

Acetylation is also promoted as cells move into S phase (for early references see Elgin & Weintraub, 1975). The use of sodium butyrate to inhibit de-acetylation has allowed more detailed examination of the time course of the modification (Sealy & Chalkley, 1979; d'Anna et al., 1980a,b). Histone H4 has been the most studied. Its primary acetylation occurs in the cytoplasm very soon after the protein has been synthesized. Further acetylations and phosphorylations then occur in the nucleus, where new histones seem to be more accessible to acetylases in vivo than those of mature nucleosomes (Cousens & Alberts, 1982), perhaps significantly contributing to the increased ^{3}H-acetate uptake observed in S phase. Regulation through the diminution of conformational constraints in S phase contrasts with regulation due to increased deacetylase activity in *Physarum* in mitosis, leading to minimal acetylation of histone H4 at metaphase (Waterborg & Matthews, 1982).

8.4.2. Histone Biosynthesis

THE MAINTENANCE OF THE DIFFERENTIATED STATE

The differentiated state must persist through S phase: newly replicated DNA must be appropriately modified and packaged so that its transcriptional pattern is identical to that in the parent cell. Experiments from Weintraub's group, with SV40, indicated that histone octamers segregated conservatively, that is, preexisting histones tended to remain together through the replicative process (Leffak et al., 1977). Further, the group demonstrated that the parental histones protected the leading side of the replicative fork—the side on which continuous DNA synthesis occurs. SV40 DNA has been completely sequenced, and it is known that the early gene products which are expressed are transcribed from the leading strand of the DNA (see Seidman et al., 1979). Conservation of histone octamers and their selective association with a particular strand of DNA suggested SV40 was an excellent model with which to study the maintenance of the differentiated state. The interpretation has been questioned, however, because of the use of cycloheximide to inhibit histone synthesis but allow already initiated DNA synthesis to be completed. Cycloheximide delays maturation of nucleosomes and causes the appearance of chromatin, which is abnormally sensitive to nuclease attack.

The distribution of newly synthesized histones is being intensively examined in more complex systems. These have many replicatory origins, and it is difficult at present to assign particular transcriptional activities among these or to the leading or lagging strands of DNA. Results with rat Morris hepatoma-F (HTC) (Jackson & Chalkley, 1981), CHO (Fowler et al., 1982), and HeLa cells (Annunziato et al., 1982) are inevitably much less clear-cut than with SV40. About 50% of newly synthesized histones H3 and H4 were deposited together on new DNA, but the remainder, and significant amounts of new histones H2A and H2B, were associated with preexisting chromatin. Russev and Hancock (1982), with mouse P815 cells, followed histone synthesis through only a short period (20%) of the S phase, minimizing potential problems from asynchrony or artefactual redistribution of nucleosomes because of the experimental conditions. Some 74% of the new histones were in nucleosomes containing newly replicated DNA. The authors cogently observed that only a small population of preexisting nucleosomes would need to remain in position to allow the transcriptional characteristics of the chromatin to be transmitted (cf. discussion on nucleosome phasing in Section 8.1.1). Leffak (1983), reported further density-labelling experiments with MSB-1 cells, which confirmed conservative assembly of histone octamer cores in nucleosomes.

HISTONE BIOSYNTHESIS

We have seen how rRNA and mRNA for many of the constitutive proteins of the cytoplasm are regulated posttranscriptionally and that facultative gene expression can be elicited in G_1 phase by transcriptionally dependent mechanisms which have not yet been clarified. The periodicity of histone biosynthesis and its coupling to DNA replication was established between 1967 and 1973, principally using HeLa and CHO cells synchronized by mitotic detachment (Robbins & Borun, 1967; Gallwitz & Mueller, 1969; Adesnick & Darnell, 1972; Gurley et al., 1972; Butler & Mueller, 1973). These classical experiments showed histones were synthesized only in S phase and were subsequently stable and that their 8–9 S messengers were not 3'polyadenylated and were present on the polysomes only in S phase. If DNA replication was blocked by hydroxyurea (inhibiting ribonucleoside diphosphate reductase) or cytosine arabinoside (inhibiting DNA polymerase), histone biosynthesis immediately ceased and the mRNA rapidly disappeared. Since that time two main questions have predominated: how is the coupling of histone and DNA synthesis achieved, and to what extent is the coupling absolute?

The Control of Histone Synthesis. This analysis became possible after some histone gene sequences had been identified, sequenced, and cloned (see Kedes, 1979; Hentschel & Birnstiel, 1981). The cDNA for sea urchin histone mRNAs was the first to be available and was used in HeLa cells to investigate the periods in the cell cycle when histone mRNA was synthesized (Melli et al., 1977). The use of heterologous cDNA can be justified because of the highly conserved sequences of the core histones, expecially histones H3 and H4; it is becoming clear, however, that this does not necessarily entail identity of the mRNAs (see Hentschel & Birnstiel, 1981; Lichtler et al., 1982). Melli and colleagues found that histone mRNA was detectable in the cytoplasm only in S phase; at other periods histone mRNA was 5–7% of that found in S phase. This was confirmed by Wilkes and colleagues

(1978). Melli and colleagues also detected histone mRNA in HeLa nuclei all through the cell cycle—an observation which was not confirmed when homologous (human) cDNA was used (Rickles et al., 1982).

Comparable studies are in hand with yeast *(Cerevisiae)*. The yeast haploid genome has two dispersed copies of the genes for histones H2A and H2B. With the use of synchronized yeast cells, mRNAs for histones 2A plus 2B were detected at the beginning of S phase and reached a peak in amount before the peak in DNA (Hereford et al., 1981). Transcription was not detected other than in S phase.

Because of our growing knowledge of the yeast genome, it is also being used to study the tight coupling between DNA and histone synthesis. The immediate switch off in transcription, as replication is blocked, is seen only for histone genes and not for others in their close proximity (Hereford et al., 1981). If an additional pair of histone H2A and H2B genes was introduced, there was no effect on yeast cell growth, and the increase in transcription of H2A-H2B mRNA was accompanied by its increased degradation so that normal steady-state levels of histone mRNA in haploid yeast were maintained (Osley & Hereford, 1981). The authors postulated that free histones might act as autogenous regulators of their own mRNA, as has been found for some ribosomal proteins in *E. coli* (Fallon et al., 1979).

The initiation of histone gene transcription in yeast at the start of S phase may be linked with the presence of autonomously replicating sequences (ars) at the 3' end of the histone H2B gene (Hereford et al., 1982). Sequences similar to ars occur in other eukaryotes (Osley & Hereford, 1982). Their relation to replicatory origins is still not clarified, and it remains to be demonstrated if histone genes are universally situated close to replicatory origins so that they are obligatorily exposed, and made available for transcription, just before or just after the onset of replication.

It is fascinating to speculate on the connection between the experiments just described and the observations of Stein and colleagues (1975) with HeLa cells, that transcription of histone mRNA genes in reconstituted chromatin could be switched on by nonhistone chromosomal proteins found in nuclei only is S phase.

Histone Variants.   Many of the early inhibition studies with hydroxyurea had indicated some continuing histone synthesis, especially of H1, after DNA replication had been blocked (Stahl & Gallwitz, 1977; Nadeau et al., 1978; Zlatanova, 1980; Chiu & Marzluff, 1982; Klenow, 1982; Shepard et al., 1982). From experiments with CHO cells held at the G_1/S border by isoleucine deprivation, Gurley and colleagues (1972) raised the possibility that some histone turnover did occur, especially with histone H1. This was confirmed in normal rat liver for highly modified species of histones H1 and H3 (Ord & Stocken, 1975b). With the development of sensitive techniques for separating closely related histones, it is now clear that some of the apparent turnover is due to the continuing synthesis of usually minor variants of histones, uncoupled to DNA replication.

The best-known of these variations is histone H5, which is found specifically in erythrocytes in all stages of their development. Histones H5 and H1 are more similar to each other than to the core histones, having flexible N and C terminal regions, the latter being relatively basic, with an apolar globular central domain (Cary et al., 1981). Subclasses of histone H1 were first recognized by Cole in the 1960s, and in 1969 Panyim and Chalkley detected a basic protein running just ahead of histone H1 on extended poly-acrylamide-urea gel electrophoresis. This $H1^0$ is now known to exist in at least two subclasses (Smith & Johns, 1980). $H1^0$ is very similar to the eythrocytic histone H5, with 70% homology in the central domain (Cary et al., 1981). The proportion of histone $H1^0$ to histone H1 increases as the mitotic activity of cells declines (Panyim & Chalkley 1969). Histones H1 and H5 are known to bind onto sites on the nucleosomes to seal off two turns of DNA, and it is expected that $H1^0$ behaves similarly. Like histone H1, $H1^0$ is phosphorylated through the cell cycle. Variants in core histones are also known (see Frankin & Zweidler, 1977 and Table 8.3). All variants

TABLE 8.3.   NONALLELIC VARIANTS IN MAMMALIAN HISTONES

Histone			
H2A-1	Thr 16	Leu 51	
H2A-2	Ser 16	Met 51	
H2B-1	Gly 75	Glu 76	
H2B-2	Ser 75	Glu 76 (so far only mouse)	
H2B-3	Gly 75	Gln 76	
H3-1	Val 89	Met 90	Cys 96
H3-2	Val 89	Met 90	Ser 96
H3-3	(Ile, Gly)		Ser 96

Source: Laine et al., 1976; Frankin & Zweidler, 1977.

are found simultaneously in all tissues in all mammals examined, but in varying proportions.

Ever since the recognition of histone H5 as a marker for nucleated erythrocytes, there has been a tantalizing possibility that some of the variations in histone type might be associated with the differentiated properties of different cell types. Two lines of evidence are beginning to suggest this may be so. First, there are the changes in proportions of histone variants with development, best-known in sea urchin embryos (see Kedes, 1979) but also well-established for H1 subtypes in a wide range of cells (Lennox & Cohen, 1983). In erythroleukemic cells after induction with DMSO or hexamethylene bisacetamide (HMBA), amounts of histone $H1^0$ increase (Kepple et al., 1977). Conversely, in pancreas regenerating after ethionine administration (Marsh & Fitzgerald, 1973) and in regenerating liver (Gjerser et al., 1982), amounts of $H1^0$ fall.

The other indication of specific roles for these histone variants comes from studies of their synthesis and modifications through the cell cycle. In CHO cells two minor variants of histone H2A (H2A-X and H2A-Z) are distinguishable from histones H2A-1 and H2A-2 by their hydroxyurea-insensitive synthesis in G_1 and G_2 as well as in S phase (Wu & Bonner, 1981; Wu et al., 1982). In synchronized HeLa cells, quite marked differences in phosphorylation patterns between histone H1 subtypes suggested that H1A and B discharge different functions; the subtype specific phosphorylations which were detected, especially on H1B, could occur in some restricted regions of the chromatin, perhaps associated with higher orders of its conformation (Ajiro et al., 1981a,b).

We do not yet know how the properties of chromatin are affected by the inclusion of a histone variant, and while there is great interest at present in the relation between nucleosome ordering and expression from the genome, much more work is required to link the presence of a particular variant to a particular requirement in chromatin function or chromosome structure and to establish how its insertion is controlled.

9
PERSPECTIVES
AND SUMMARY

9.1. THE REGENERATIVE CAPACITY IN VERTEBRATES

Survival of the organism following deprivation of a part may require replacement by a replica of the original structure, as with a limb, but regeneration of an organ has only to satisfy the functional needs of the organism
and may be met by hypertrophy or hyperplasia, as seen in liver (Section
3.3), pancreas (see Fitzgerald, 1980), and salivary gland (Baserga, 1976).
In assessing the obvious limitations shown in the capacity of animals to
regenerate, we can distinguish cells and tissues which never regenerate in
any vertebrate—notably neurons—from cases in which regeneration can
occur in some orders—the urodela—but not in others within the same class—
the anura.

9.1.1. Cells with Inherent Incompatibilities between Their Function and Replacements (see Ham & Veomett, 1980)

NEURONS

It is usually thought that the elaborate and extensive intercellular connections
in which neurons, especially of the central nervous system, are involved
necessitate the maintenance of the neuron in the G_0 state. If cells enter a
growth cycle and pass through mitosis, major changes in the plasma membrane
follow (Chapter 6), which would jeopardize the essential continuity of the
synapses, even though structural components within a synapse are continuously being turned over. How neuronal reentry into the growth cycle is
irreversibly blocked under normal circumstances is unknown. Most simply,
a component unique to terminally differentiated neurons could prevent an
essential step in the transit between the G_0 and the G_1 states.

HEMOPOIETIC TISSUE

The pros and cons of the various differentiational strategies so far distinguishable (Section 8.2.3) are speculative. If partial or complete elimination
of DNA sequences is involved, as with cells in the lymphoid and mammalian
erythroid series, progenitor (stem) cells, which have not yet undergone the
irreversible step in the differentiation process, must be maintained if tissue
depletion is to be repairable.

OTHER EPITHELIA

Cells of the intestinal epithelium and skin are continuously and rapidly replaced. As with neurons, the terminal differentiated state is physiologically, but not experimentally (Section 1.1.2), irreversible. The extracellular environment of the terminally differentiated cells in vivo is markedly different from that into which nuclei have been transplanted in Gurdon's classical experiments with *Xenopus* ova (Section 1.1.2). With keratinocytes, diminishing access of nutrients and growth promoters is probably the limiting factor causing a switch from proliferation to the G_0 state, but this is improbable for the intestinal epithelium. Here, however, heavy demands for ATP by the actively transporting cells may cause their energy status to be precariously balanced.

9.1.2. Strategies in Gene Location

Three assumptions about the bases of differentiation have been implicit in the analysis of how cells in G_0 may be promoted into growth:

1. Gene subsets within the total genome, which may be required to be expressed during the lifetime of the determined cell or its progeny, are organized so that they can be available for transcription, that is, they must not be located in or near (Section 8.1.4) obligatorily heterochromatic regions. If the gene products are not continuously required, regulatory sites must be potentially accessible to signals from the cytoplasm.
2. The differentiated state is determined early in embryogenesis.
3. The process by which gene transcription is determined has inherently associated with it the machinery by which active chromatin is distinguishable, for example, increased acetylation of core histones (Section 8.1.3).

9.1.3. The Pleiotropic Response to Growth Stimulation

That it is inherent for cells to respond pleiotropically to hormone stimulation was first enunciated by Tomkins and his co-workers (Hershko et al., 1971). What has emerged since, about both the behavior of the cell membrane in response to ligand binding (Section 6.1) and the interacting and cascading reactions which may follow intracellularly (Chapters 5 & 7), serves to reinforce Tomkins' concept. For animal cells, growth promotion requires adequate nutrient uptake and its efficient utilization. Increased ATP turnover ensues, and a more efficient conversion of pre-mRNA transcripts to mRNA and its increased translation follows.

Each of these can follow from the binding of a cAMP-releasing hormone and/or insulin (Chapter 4; see Table 9.1). As such binding occurs normally during homeostatic regulation in intact animals without enhancing growth, tissue-specific growth-inhibitory mechanisms are usually postulated, also acting on the plasma membrane (Section 4.2.2). Tissue depletion would cause the amount of inhibitor to fall, and this release of membrane constraints then permits cells to move out of the G_0 state.

9.1.4. Mitogenesis

Conditions required for cells to proceed from growth early in G_1 to replication, which is usually, though not invariably (e.g., polyploidy in liver parenchymal cells) followed in animal cells by mitosis and cytokinesis, are a major area of ignorance in cell biology. For vertebrate cells in culture, the transit can often be specifically promoted so that numbers of mitoses exceed those found in the presence of nonlimiting nutrients, insulin, and an adequate but inert substratum. Complications may arise analytically, since it is evident from cell cultures that proliferating cells commonly release auto-regulatory materials affecting their own growth (Section 4.2.2).

Currently, a process potentially critical for transition from growth to replication is tyrosinyl protein phosphorylation of an unknown substrate (Section 4.2.2), and this process could be directly or indirectly counteracted by growth inhibitors. With cells in culture, EGF, PDGF, IGF-2, and insulin and IGF-1 through their interactions with IGF-2 and its receptors, are established mitogens. We do not yet know what the factors are which promote proliferation in the blastema, but as conditions for culturing blastemal cells become established, sensitive techniques are now available for determining whether prostaglandins, thrombin, and/or peptides, with some homologies to those released in higher vertebrates, are involved.

9.1.5. Tissue Reconstruction

Mitogenesis has to be followed by positionally defined specification of differentiated cell types. It is clear (Section 2.2.3) that, for symmetrically

TABLE 9.1. CHRONOLOGY OF PLEIOTROPIC ACTIONS OF GROWTH FACTORS

Response of Cultured Cells to EGF	Time	General Response of Cells to Growth Factors in Vivo	Time
Initiation EGF Binding Clustering bound EGF Phosphorylation membrane proteins	0 h	Growth Factor(s) Bound Increased Na$^+$ entry Intracellular pH increases Increased K$^+$ transport	0 h
Maximal membrane ruffling	3 min		
Internalization bound EGF	6 min		
Appearance bound EGF in lysosomes	0.5 h	Increased amino acid transport Increased pyrimidine riboside transport	0.5 h
Completion EGF binding Maximal EGF degradation Maximal EGF receptor degradation Maximal EGF receptor down-regulation	1 h	Increased glycolysis	1 h
	2 h	Maximum increase in glucose transport Increased cyclic AMP detectable Increase in ornithine decarboxylation Increase in protein synthesis	2 h
	5 h	Increased rRNA in cytoplasm Increased met tRNA$_f$ in cytoplasm	5 h
	10 h		10 h

Source: The responses of cultured cells to EGF are taken from Fox et al., 1982. General pleiotropic responses of cells to growth factors are from Fox et al., 1982, and Chapters 4–8.

179

organized structures, determination of AP and DV symmetry occurs very early and is effectively irreversible, persisting in and being transmitted by cells in the amputation plane.

Recent results from Stocum (1982), Slack (1983), and others support skin (wound epidermis and/or dermis, MacCabe et al., 1974) as a major positional influence in restructuring the AP and DV axes of the regenerating limb.

Phenotypic characteristics distinguishing epidermal cells which convey AP rather than DV positional information have not yet been identified and little is known of the determinants of the proximo-distal axis in vertebrates, although in invertebrates a metabolic parameter has been indicated (Section 1.3.1).

What is communicated by the cells has not yet been established. Embryological experiments were classically interpreted to indicate "instructive" and "permissive" information (Holtzer et al., 1975; Wessells, 1977). With greater knowledge of ligand-receptor interactions at the cell membrane (Section 6.1), of processes of signal amplification (Chapters 5 & 6), and with slightly more information about potential mechanisms for regulating phenotypic expression in animal cells (Chapter 8), it may now be appropriate to consider growth-regulatory and positional information exchanged between cells in the light of the ways in which the signals are received and transduced by the cells. Two classes of stimulation can be distinguished, the first biochemical, exemplified in the release by cells of growth factors which interact with recipient cells through specific high-affinity receptors. Amplified intracellular responses are produced, which in some cases have localized structural effects because of the interactions between clustered growth-factor receptor complexes and the cytoskeleton. The second, biophysical class of stimuli affects recipient cells without the need for specific high-affinity receptors, so the physical properties of the membrane are altered, and the cells round up, or flatten, or the lateral mobility of their peripheral constituents is constrained (Chapter 6). This class is illustrated by the effects that contact with components of the extracellular matrix has on the differentiated expression of cells (see Wessells, 1977; Bissell et al., 1982); for limb regeneration the relevant extracellular components would be synthesized and secreted by cells of the epidermis and/or dermis.

The immediate consequences of biophysical stimulation of the plasma membrane may be indistinguishable from biochemical effects, in that changes in transport, ion fluxes (Section 5.2.2), and cytoskeletal responses (Chapter

6) are all pleiotropic accompaniments of specific growth-factor stimulation. There is, however, evidence that interaction through specific receptors can produce discriminatory responses. In adrenal cortical cells, low levels of ACTH elicit maximum steroidogenesis, although with increasing levels of the hormone, amounts of cAMP continue to rise (Section 4.1).

Phosphorylation status, redox change, and intracellular pH have been advanced as the quasi-independent variables which are altered by the intercellular growth-regulatory signals so far known. If tyrosine phosphorylation proves to be critical in moving a cell from G_1 into S phase, we still do not know the biochemical requirements for exit from the growth cycle into the G_0 state nor for the expression in the G_0 state of differentiated characteristics. We have also to explain how the same agent, for example, cAMP, can facilitate growth and mitosis and keep cells quiescent. The range of specificity of seryl protein kinases, along with the relatively large number (up to 10%) of seryl residues which frequently occurs in proteins, raises the possibility (Section 7.4) that seryl protein phosphorylation is one of the devices which cells employ to change from analogue signaling at the plasma membrane and in the cytoplasm to digital switching in the nucleus.

9.1.6. Limitation to Regeneration

The limitations to regeneration must be compatible with any picture we have of its normal mechanism. There are several requirements for limb regeneration to occur. Cells must be available to form the blastema and must subsequently be sustained in their activities by appropriate nutrients and the maintenance hormones on which their normal metabolism depends. Once the blastema is established, a proper balance is needed between proliferation, differentiation, and tissue restructuring.

Two features are usually stressed: innervation and the characteristics of the wound epidermis (see Wallace, 1981; Tassava & Olsen, 1982). Many classical experiments demonstrate the importance of innervation in the establishment of the blastema in limb regeneration and of the quantitative relation between nerve supply and the potentially regenerating site (but see Section 2.1.2). The basis of the neurotrophic effect is still unknown, and the release of a mitogen of neuronal origin remains to be established.

For compensatory hyperplasia following cell or tissue depletion in higher vertebrates, nerve involvement is unproven or nonessential. Tassava and Olsen (1982) stress the importance of the wound epidermis in preventing premature differentiation. When scar tissue covers an amputation stump, or if the blastemal cells lose contact with the epidermis, as when whole skin (epidermis and dermis) is made to cover the stump, regeneration is prevented. The epidermis has previously been implicated as a potential source of positional information (Section 9.1.5). Its basement membrane may be a source of mitogen (Greenburg & Gospodarowicz, 1982), and continued, localized intercellular signals with feedback may be the means by which positional information is conveyed. Alteration in the balance of glycopeptides adjacent to the epidermis, such as fibronectin and hyaluronic acid, have already been described in newt limb as regeneration proceeds (Repesh et al., 1982; Donaldson & Mahan, 1983). Different proportions of the components underlying scar tissue may explain the inability of the latter to sustain regeneration.

A multiplicity of biochemical changes have been described which take place during the progress of a cell from the G_0 state to mitosis. Release of secondary messengers, changes in amino acid and ion transport, increased transcription, the induction of enzymes needed for the synthesis of DNA are all well established. Unfortunately these data only present a composite picture drawn from various cell types; in no case is a complete and detailed picture available for a single class of cells. The results do not tell us what biochemical information in the otherwise intact animal instructs the cell to divert from its replicatory activities into a stationary phase or to adopt a particular phenotypic expression when it differentiates, as required in regeneration. It is hoped that a coalescence of diverse disciplines—genetics, experimental embryology and ingenious operative procedures, morphological observations, immunochemistry, and molecular biology, together with the sophisticated biochemical techniques now available—can soon be applied to give a molecular interpretation to the regenerative process.

BIBLIOGRAPHY

Abraham, K. A., I. F. Pryme, A. Åbro, & R. M. Dowben, *Exp. Cell Res. 82*, 95–102 (1973).

Acher, R., *Proc. Roy. Soc. London Ser. B 210*, 21–43 (1980).

Adamietz, P., K. Wielckens, R. Bredehorst, H. Lengyel, & H. Hilz, *Biochem. Biophys. Res. Commun. 101*, 96–103 (1981).

Adesnick, M., & J. E. Darnell, *J. Mol. Biol. 67*, 397–406 (1972).

Ajiro, K., T. W. Borun, & L. H. Cohen, *Biochemistry 20*, 1445–1454 (1981a).

Ajiro, K., T. W. Borun, S. D. Shulman, G. M. McFadden, & L. H. Cohen, *Biochemistry 20*, 1454–1464 (1981b).

Ali, I. U., & R. O. Hynes, *Cell 14*, 439–446 (1978).

Allen, J. C., & C. J. Smith, *Biochem. Soc. Trans. 7*, 584–592 (1979).

Allis, C. D., & M. A. Gorovsky, *Biochemistry 20*, 3828–3833 (1981).

Amara, S. G., V. Jonas, M. G. Rosenfeld, E. S. Ong, & R. M. Evans, *Nature (Lond.) 298*, 240–244 (1982).

Anderson, R. E., & J. C. Standefer, in *Cytotoxic Insult to Tissue*, C. S. Potten & J. H. Hendry, eds., Churchill Livingstone, New York, pp. 67–104 (1982).

Anderson, R. E., & N. L. Warner, *Adv. Immunol. 24*, 215–335 (1976).

Andreis, P. G., J. F. Whitfield, & U. Armato, *Exp. Cell Res. 134*, 265–272 (1981).

d'Anna, J. A., L. R. Gurley, R. R. Becker, S. S. Barham, R. A. Tobey, & R. A. Walters, *Biochemistry 19*, 4331–4341 (1980b).

d'Anna, J. A., R. A. Tobey, & L. R. Gurley, *Biochemistry 19*, 2656–2671 (1980a).

Annunziato, A. T., R. K. Schindler, M. G. Riggs, & R. L. Seale, *J. Biol. Chem. 257*, 8507–8515 (1982).

Arch, J. R. S., & E. A. Newsholme, *Essays in Biochemistry 14*, 82–123 (1978).

Aristotle (330 BC[a]), *De Partibus Animalium, Book II* (W. Ogle, trans.—Works of Aristotle), J. A. Smith & W. D. Ross, eds., Clarendon Press, Oxford, p. 508 (1910).

Aristotle (330 BC[b]), *Historia Animalium, Book IV* (D'Arcy Thompson, trans.), J. A. Smith & W. D. Ross, eds., Clarendon Press, Oxford, p. 771d (1910).

Aschinberg, L. C., G. Koskimies, J. Bernstein, M. Nash, C. M. Edelmann, & A. Spitzer, *Yale J. Biol. Med. 51*, 341–345 (1978).

Atkinson, D., *Biochemistry 7*, 4030–4034 (1968).

Atmar, V. J., & G. D. Kuehn, *Proc. Natl. Acad. Sci. 78*, 5518–5522 (1981).

Atryzek, V., & N. Fausto, *Biochemistry 18*, 1281–1287 (1979).

Austin, S. A., & M. J. Clemens, *Biosci. Rep. 1*, 35–44 (1981).

Austin, S. A., & J. E. Kay, *Essays in Biochemistry 18*, 79–120 (1982).

Autuori, F., P. Baldini, A. Ciofiluzzatto, L. C. Devirgiliis, L. Dini, S. Incerpi, & P. Luly, *Biochim. Biophys. Acta 678*, 1–6 (1981).

Azarnia, R., G. Dahl, & W. R. Loewenstein, *J. Membr. Biol. 63*, 133–146 (1981).

Bakke, O., & O. W. Rønning, *J. Cell Physiol. 113*, 459–464 (1982).

Baldwin, G. S., J. Knesel, & J. M. Monckton, *Nature (Lond.) 301*, 435–437 (1983).

Ballal, N. R., Y. J. Kang, M. O. J. Olson, & H. Busch, *J. Biol. Chem. 250*, 5921–5925 (1975).

Ballard, F. J., & G. L. Francis, *Biochem. J. 210*, 243–249 (1983).

Banchereau, J. F. J., D. M. Danois, M. Guenounou, G. M. S. Durand, & J. C. L. Agneray, *Biochim. Biophys. Acta 678*, 98–105 (1981).

Barbiroli, B., & V. R. Potter, *Science 172*, 738–741 (1971).

Barlow, S. D., *Biochem. J. 154*, 395–403 (1976).

Barlow, S. D., & M. G. Ord, *Biochem. J. 148*, 295–302 (1975).

Barnes, D. W. H., C. E. Ford, S. M. Gray, & J. F. Loutit, *Prog. Nucl. Energy (Biol) 2*, 1–10 (1959).

Barondes, S. H., *Annu. Rev. Biochem. 50*, 207–231 (1981).

Barrack, E. R., & D. S. Coffey, *Recent Prog. Horm. Res. 38*, 133–189 (1982).

Bar-Shavit, R., A. Kahn, J. W. Fenton, & G. D. Wilner, *J. Cell Biol. 96*, 282–285 (1983).

Barth, L., *Biol. Bull. Woods Hole 74*, 155–177 (1938).

Baserga, R., *Multiplication and Division in Mammalian Cells*, Marcel Dekker, New York (1976).

Bauer, F. C. H., & M. R. Urist, *Clin. Orthop. 154*, 291–295 (1981).

Baydoun, H., J. Hoppe, & K. G. Wagner, *Cold Spring Harbor Conferences on Cell Proliferation 8*, 1095–1108 (1981).

Beach, D., B. Durkacz, & P. Nurse, *Nature (Lond.) 300*, 706–709 (1982).

Beck, W. T., R. A. Bellantone, & E. S. Canellakis, *Biochem. Biophys. Res. Commun. 48*, 1649–1655 (1972).

Beis, I., & E. A. Newsholme, *Biochem. J. 152*, 23–32 (1975).

Bengtsson, B. G., K. H. Kiessling, A Smith-Kielland, & J. Morland, *Eur. J. Biochem. 118*, 591–597 (1981).

Benya, P. D., & J. D. Shaffer, *Cell 30*, 215–224 (1982).

Berking, S., *Wilhelm Roux's Arch. Entwicklungsmech Org. 141*, 99–110 (1977).

Berlin, R. D., & J. M. Oliver, *J. Cell Biol. 77*, 789–804 (1978).

Berlin, R. D., & J. M. Oliver, *J. Theor. Biol. 99*, 69–80 (1982).

Berridge, M., *Adv. Cyclic Nucleotide Res. 6*, 1–98 (1975).

Berridge, M. J., in *Insect Biology in the Future*, V. B. Wigglesworth & M. Locke, eds., Academic Press, New York, pp. 463–478 (1980).

Besterman, J. M., & R. B. Low, *Biochem. J. 210*, 1–13 (1983).

Beyer, A. L., M. E. Christensen, B. W. Walker, & W. M. le Stourgeon, *Cell 11*, 127–138 (1977).

Bhorjee, J. S., *Proc. Natl. Acad. Sci. 78*, 6944–6948 (1981).

Bhorjee, J. S., I. Mellon, & L. Kifle, *Biochem. Biophys. Res. Commun. 111*, 1001–1007 (1983).

Bintliff, S., & B. E. Walker, *Am. J. Anat. 106*, 233–245 (1960).

Birchmeier, W., *Trends in Biochem. Sc. 6*, 234–237 (1981).

Bird, A. P., *J. Mol. Biol. 118*, 49–60 (1978).

Bissell, M. J., H. G. Hall, & G. Parry, *J. Theor. Biol. 99*, 31–68 (1982).

Blazquez, E., B. Ribalcava, R. Montesano, L. Orci, & R. H. Ungar, *Endocrinology 98*, 1014–1023 (1976).

Bloom, K. S., & J. N. Anderson, *Cell 15*, 141–150 (1978).

Blumberg, P. M., & P. W. Robbins, *Cell 6,* 137–147 (1975).

Bonghetti, A. F., J. A. Kay, & K. P. W. Lealer, *Biochem, J. 182,* 27–32 (1978).

Bonner, W. M., *J. Cell Biol. 64,* 421–430, 431–437 (1975).

Boogaard, C., & G. H. Dixon, *Exp. Cell Res. 143,* 191–205 (1983).

Boonstra, J., C. L. Mummery, L. G. J. Terttoolen, P. T. van der Saag, & S. W. de Laat, *J. Cell Physiol. 107,* 75–83 (1981).

Borgens, R. B., *Int. Rev. Cytol. 76,* 245–298 (1982).

Bourne, G. H., in *Tissue Repair and Regeneration,* L. E. Glynn, ed., Elsevier North-Holland, New York, pp. 211–243 (1981).

Bowen, B. C., *Nucleic Acids Res. 9,* 5093–5108 (1981).

Boynton, A. L., & J. F. Whitfield, *Exp. Cell Res. 126,* 477–481 (1980).

Boynton, A. L., J. F. Whitfield, J. P. Macmanus, U. Armato, B. K. Tsang, & A. Jones, *Exp. Cell Res. 135,* 199–211 (1981).

Braciale, V. L., J. R. Gavin & T. J. Braciale, *J. Exp. Med. 156,* 664–669 (1982).

Bradbury, E. M., N. Maclean, & H. R. Matthews, *DNA, Chromatin and Chromosomes,* Blackwell Scientific Publications, Boston (1981).

Braun, J., K. Fujiwara, K. T. D. Pollard, & E. R. Unanue, *J. Cell Biol. 79,* 409–418 (1978).

Braun, S., A. M. Tolkovsky, M. L. Steer, H. A. Lester, & A. R. Levitski, *Biochem. Soc. Trans. 10,* 496–498 (1982).

Breathnach, R., C. Benoist, K. O'Hare, F. Gannon, & P. Chambon, *Proc. Natl. Acad. Sci. 75,* 4853–4857 (1978).

Breathnach, R., & P. Chambon, *Annu. Rev. Biochem. 50,* 349–383 (1981).

Bredehorst, R., K. Wielckens, P. Adamietz, E. Steinhagen-Thiessen, & H. Hilz, *Biochemistry 120,* 267–274 (1981).

Bresnick, E., *Methods Cancer Res. 6,* 347–397 (1971).

Bretscher, M. S., *Nature (Lond.) 260,* 21–23 (1976).

Brønsted, H. V., *Planarium Regeneration,* Pergamon Press, Oxford (1969).

Brønsted, G., & T. Christoffersen, *FEBS Lett. 12,* 89–93 (1980).

Brooks, R. F., in *The Cell Cycle,* P. C. L. John, ed., Cambridge University Press, New York, pp. 35–61 (1981).

Brooks, R. F., D. C. Bennett, & J. A. Smith, *Cell 19,* 493–504 (1980).

Brown, R. L., R. L. Griffith, R. H. Neubauer, & H. Rabin, *J. Cell Physiol. 115,* 191–198 (1983).

Bryan, P. N., H. Hofstetter, & M. L. Birnstiel, *Cell 27,* 459–466 (1981).

Bryant, S. V., *Nature (Lond.) 263,* 676–679 (1976).

Bryant, S. V., V. French, & P. J. Bryant, *Science 212,* 993–1002 (1981).

Bryant, S. V., & L. E. Iten, *Dev. Biol. 50,* 212–234 (1976).

Bucher, N. L. R., & R. A. Malt, *Regeneration of Liver and Kidney,* Little, Brown & Co., Boston (1971).

Bucher, N. L. R., & M. W. Swaffield, *Adv. Enzyme Regul. 13,* 291–293 (1976).

Bullough, W. S., *Cancer Res. 25,* 1683–1727 (1965).

Bullough, W. S., & E. B. Lawrence, *Proc. Roy. Soc. London Ser. B 151,* 517–536 (1959–1960).

Burgoyne, R. D., *FEBS Lett. 94*, 17–19 (1978).

Burns, F. J., & I. F. Tannock, *Cell Tissue Kinet. 3*, 321 (1970).

Burnstock, G., *Pharmacol. Rev. 24*, 509–581 (1972).

Butler, E. G., *J. Exp. Zool. 65*, 271–316 (1933).

Butler, W. G., & G. C. Mueller, *Biochim. Biophys. Acta 294*, 481–496 (1973).

Butt, T. R., J. Brothers, C. P. Giri, & M. E. Smulson, *Nucleic Acids Res. 5*, 2775–2788 (1978).

Campisi, J., E. E. Medrano, G. Morreo, & A. B. Pardee, *Proc. Natl. Acad. Sci. 79*, 436–440 (1982).

Candido, E. P. M., R. Reeves, & J. R. Davie, *Cell 14*, 105–113 (1978).

Canellakis, E. S., J. J. Jaffe, R. Mantsavinos, & J. S. Krakow, *J. Biol. Chem. 234*, 2096–2099 (1959).

Caplan, A. I., in *Levels of Genetic Control in Development*, S. Subtelny & U. K. Abbott, eds., Alan R. Liss, Inc., New York, pp. 37–68 (1981).

Caplan, A., M. G. Ord, & L. A. Stocken, *Biochem J. 174*, 475–483 (1978).

Carpenter, G., & S. Cohen, *Annu. Rev. Biochem. 48*, 193–216 (1979).

Carroll, J. M., & R. E. Sicard, *Wilhelm Roux's Arch. of Dev. Biol. 189*, 107–110 (1980).

Carter, B. L. A., in *The Cell Cycle*, P. C. L. John, ed., Cambridge University Press, Cambridge, England, pp. 99–117 (1981).

Cartwright, I. L., M. A. Keene, G. C. Howard, S. M, Abmayr, G. Fleischmann, K. Lowenhaupt, & S. C. R. Elgin, *Crit. Rev. Biochem. 13*, 1–86 (1982).

Cary, P. D., M. L. Hines, E. M. Bradbury, B. J. Smith & E. W. Johns, *Eur. J. Biochem. 120*, 371–377 (1981).

Cato, A. C. B., *Biosci. Rep. 3*, 101–111 (1983).

Chafouleas, J. G., W. E. Bolton, H. Hidaka, A. E. Boyd, & A. R. Means, *Cell 28*, 41–50 (1982).

Chambard, J. C., A. Franchi, A. LeCam, & J. Pouyssegur, *J. Biol. Chem. 258*, 1706–1713 (1983).

Chan, L. N., & W. Gehring, *Proc. Natl. Acad. Sci. 68*, 2217–2221 (1971).

Chang, K.-J., & P. Cuatrecasas, *J. Biol. Chem. 249*, 3170–3180 (1974).

Chen, C. C., B. B. Bruegger, C. W. Kern, Y. C. Lin, R. M. Halpern, & R. A. Smith, *Biochemistry 16*, 4852–4855 (1977).

Chen, H. W., *J. Cell. Physiol. 108*, 91–97 (1981).

Chen, H. W., H. J. Heiniger, & A. A. Kandutsch, *Proc. Natl. Acad. Sci. 72*, 1950–1954.

Chen, L. B., R. C. Gudor, T. Y. Sun, A. B. Chen, & M. W. Mosesson, *Science 197*, 776–778 (1977).

Cheung, W. Y., *Science 207*, 19–27 (1980).

Child, C. M., *Patterns and Problems of Development*, University of Chicago Press, Chicago, Illinois (1941).

Chiquet, M., H. M. Eppenberger, & D. C. Turner, *Dev. Biol. 88*, 220–235 (1981).

Chiu, I. M., & W. F. Marzluff, *Biochim. Biophys. Acta 699*, 173–182 (1982).

Chock, P. B., S. G. Rhee, & E. R. Stadtman, *Annu. Rev. Biochem. 49*, 813–843 (1980).

Chytil, F., & D. E. Ong, *Vit. & Horm. 36*, 1–22 (1978).

Clague, M. B., in *Nitrogen Metabolism in Man*, J. C. Waterlow & J. M. L. Stephen, eds., Applied Science, Barking, England, pp. 525–539 (1982).

Clemens, M. J., & V. M. Pain, *Biochim. Biophys. Acta 361*, 345–357 (1974).

Clemens, M. J., V. M. Pain, S. T. Wong, & E. C. Henshaw, *Nature (Lond.) 296*, 93–95 (1982).

Cobbold, P. H., & M. H. Goyns, *Biosci. Rep. 3*, 79–86 (1983).

Cohen, P., in *Molecular Aspects of Cell Regulation 1*, P. Cohen, ed., Elsevier North-Holland, New York, pp. 253–268 (1980).

Cohen, P., D. Yellowlees, A. Aitken, A. D. Deana, B. A. Hemmings, & P. J. Parker, *Eur. J. Biochem. 124*, 21–35 (1982).

Cohen, S., R. A. Fava, & S. T. Sawyer, *Proc. Natl. Acad. Sci. 79*, 6237–6241 (1982).

Cohen, S., & J. M. Taylor, *Rec. Prog. Horm. Res. 30*, 533–550 (1974).

Cohen, S. S., *Nature (Lond.) 274*, 209–210 (1978).

Cook, P. R., & J. Brazell, *J. Cell Sci. 19*, 261–279 (1975).

Cook, P. R., J. Long, A. Hayday, L. Lania, M. Fried, D. J. Chiswell, & J. A. Whyke, *EMBO J. 1*, 447–452 (1982).

Cooper, A. R., M. Kurkinen, A. Taylor, & B. L. M. Hogan, *Eur. J. Biochem. 119*, 189–197 (1981).

Cooper, D. N., M. H. Taggart, & A. P. Bird, *Nucleic Acids Res. 11*, 647–658 (1983).

Cooper, H. L., & R. Braverman, *J. Biol. Chem. 256*, 7461–7467 (1981).

Cooper, H. L., M. H. Park, & J. E. Folk, *Cell 29*, 791–797 (1982).

Cooper, J. A., N. A. Reiss, R. J. Schwartz, & T. Hunter, *Nature (Lond.) 302*, 218–223 (1983).

Cornell, R. P., *Am. J. Physiol. E3*, 112–118 (1981).

Costa, M., E. W. Gerner, & D. H. Russell, *Biochim. Biophys. Acta 425*, 246–255 (1976).

Costa, M., E. W. Gerner, & D. H. Russell, *Biochim. Biophys. Acta 538*, 1–10 (1978).

Cousens, L. S., & B. M. Alberts, *J. Biol. Chem. 257*, 3945–3949 (1982).

Craddock, V. M., *Chem.-Biol. Interact. 10*, 313–323 (1975).

Creek, K. E., D. J. Morré, C. S. Silverman-Jones, Y. Shidoji, & L. M. De Luca, *Biochem. J. 210*, 541–547 (1983).

Crick, F. H. C., & P. A. Lawrence, *Science 189*, 340–347 (1975).

Cross, M. E., & M. G. Ord, *Biochem. J. 124*, 241–248 (1971).

Cummings, B., M. R. Kaser, G. Wiggins, M. G. Ord, & L. A. Stocken, *Biochem. J. 208*, 141–146 (1982).

Czech, M. *Cell 31*, 8–10 (1982).

Dabeva, M. D., & K. P. Dudov, *Biochem. J. 204*, 179–183 (1982).

Dahmus, M. E., *J. Biol. Chem. 256*, 3319–3325 (1981).

Darnell, J. E., *Nature (Lond.) 297*, 365–371 (1982).

Das, M., *Int. Rev. Cytol. 78*, 233–256 (1982).

Davidson, B. L., J. M. Egly, E. R. Mulvihill, & P. Chambon, *Nature (Lond.) 301*, 680–686 (1983).

Davidson, E. H., H. T. Jacobs, & R. J. Britten, *Nature (Lond.) 301*, 468–470 (1983).

Davie, J. R., & E. P. M. Candido, *Proc. Natl. Acad. Sci. 75*, 3574–3577 (1978).

Davies, D. J., & G. B. Ryan, in *Nutrition and Wound Healing*, L. E. Glynn, ed., Elsevier North-Holland, New York, pp. 515–574 (1981).

de Laart, S. W., & P. T. van der Saag, *Int. Rev. Cytol. 74*, 1–54 (1982).

De Luca, F., *Vit. & Horm. 35*, 1–57 (1977).

Desselle, J.-C., & M. Gontcharoff, *Biol. Cellulaire 33*, 45–54 (1978).

Deutsch, C., & M. Price, *J. Cell. Physiol. 113*, 73–79 (1982).

Dexter, T. M., T. D. Allen, & L. G. Lajtha, *J. Cell. Physiol. 91*, 335–344 (1977).

Dexter, T. M., N. G. Testa, T. D. Allen, T. Rutherford, & E. Scholnick, *Blood 58*, 699–707 (1981).

Dienstman, S. R., & H. Holtzer, in *Cell Cycle and Cell Differentiation*. J. Reiner & H. Holtzer, eds., Springer-Verlag, New York, pp. 1–25 (1975).

Dierks, P., A. van Ooyen, M. D. Cochran, C. Dobkin, J. Reiser, & C. Weissman, *Cell 32*, 695–706 (1983).

Diesterhaft, M., T. Naguchi, & D. Granner, *Eur. J. Biochem. 108*, 357–365 (1980).

Dingle, J. T., & J. A. Lucy, *Biol. Rev. 40*, 422–461 (1965).

Dingwall, C., G. F. Lomonossoff, & R. A. Laskey, *Nucleic Acids Res. 9*, 2659–2673 (1981).

Dingwall, C., S. V. Sharnick, & R. A. Laskey, *Cell 30*, 449–458 (1982).

Dolnick, B. J., R. J. Berenson, J. R. Bertino, R. J. Kaufman, J. H. Nunberg, & R. T. Schimke, *J. Cell Biol. 83*, 394–402 (1979).

Donachie, W. D., in *The Cell Cycle*, P. C. L. Johns, ed., Cambridge University Press, Cambridge, England, pp. 63–83 (1981).

Donaldson, D. J., & J. T. Mahan, *J. Cell Sci. 62*, 117–127 (1983).

Dräger, U. C., *Nature (Lond.) 303*, 169–172 (1983).

Duceman, B. W., & S. T. Jacob, *Biochem. J. 190*, 781–789 (1980).

Dudov, K. P., & M. D. Dabeva, *Biochem. J. 210*, 183–192 (1983).

Durham, A. Ch., & J. M. Walton, *Biosci. Rep. 2*, 15–30 (1982).

Durkacz, B. W., O. Omidiji, D. A. Gray, & S. Shall, *Nature (Lond.) 283*, 593–596 (1980).

Eagle, H., *Science 148*, 42–51 (1965).

Edelman, I. S., *J. Steroid Biochem. 6*, 147–149 (1975).

Egan, P. A., & B. Levy-Wilson, *Biochemistry 20*, 3695–3702 (1981).

Eisen, H. N., *Immunology*, 2d ed., Harper & Row, New York (1980).

Ek. B., B. Westmark, A. Wasteson, & C.-H. Heldin, *Nature (Lond.) 295*, 419–420 (1982).

Elgin, S. C. R., *Cell 27*, 413–415 (1981).

Elgin, S. C. R., & H. Weintraub, *Annu. Rev. Biochem. 44*, 725–774 (1975).

Emerman, J. T., J. Bartley, & M. J. Bissell, *Exp. Cell Res. 134*, 241–250 (1981).

Erslev, A. J., *Am. J. Pathol. 65*, 629–639 (1971).

Ewton, D. Z., & J. R. Florini, *Dev. Biol. 86*, 31–39 (1981).

Fallon, A. M., C. S. Jinks, M. Yamamoto, & M. Nomura, *J. Bacteriol. 138*, 383–396 (1979).

Fallon, R. F., & D. A. Goodenough, *J. Cell. Biol. 90*, 521–526 (1981).

Fantes, P. A., & P. Nurse, in *The Cell Cycle*, P. C. L. John, ed., Cambridge University Press, Cambridge, England, pp. 11–33 (1981).

Farquhar, M. G., in *Biology and Chemistry of Basement Membranes*, N. A. Kefalides, ed., Academic Press, New York, pp. 43–80 (1978).

Fausto, N., *Biochim. Biophys. Acta 238*, 116–128 (1971).

Fausto, N., & F. R. Butcher, *Biochim. Biophys. Acta 428*, 702–706 (1976).

Fehlman, M., J-L. Carpenter, E. van Obberghen, P. Freychet, P. Thamm, D. Saunders, D. Brandenburg, & L. Orcl, *Proc. Natl. Acad. Sci. 79*, 5921–5925 (1982).

Felber, S. M., & M. D. Brand, *Biochem. J. 210*, 885–891, 893–897 (1983).

Feldherr, C. M., & J. A. Ogburn, *J. Cell. Biol. 87*, 589–593 (1980).

Fell, H. B., *Embryologia 10*, 181–205 (1969).

Fell, H. B., & E. Mellanby, *J. Physiol. 116*, 320–349 (1952).

Fell, H. B., & E. Mellanby, *J. Physiol. 119*, 470–488 (1953).

Ferris, G. M., & J. B. Clark, *Biochim. Biophys. Acta 273*, 73–79 (1972).

Fialkow, P. J., *J. Cell. Physiol. Suppl. 1*, 37–43 (1982).

Fisher, J. W., *Proc. Soc. Exp. Biol. Med., 173*, 289–305 (1983).

Fitzgerald, P. J., in *The Pancreas*, P. J. Fitzgerald & A. B. Morrison, eds., Williams & Wilkins, Baltimore, Maryland, pp. 1–29 (1980).

Flagg-Newton, J. L., G. Dahl, & W. R. Loewenstein, *J. Membr. Biol. 63*, 105–21 (1981).

Flagg-Newton, J. L. & W. R. Loewenstein, *J. Membr. Biol. 63*, 123–131 (1981).

Flanagan, J., & G. Le Koch, *Nature (Lond.) 273*, 278–281 (1978).

Flickinger, R. A., *Growth 23*, 251–271 (1959).

Flickinger, R. A., *Int. Rev. Cytol. 75*, 229–241 (1982).

Flickinger, R. A., & S. J. Coward, *Dev. Biol. 5*, 179–204 (1962).

Foidart, J. M., & A. H. Reddi, *Dev. Biol. 75*, 130–136 (1980).

Folkman, J., & A. Moscona, *Nature (Lond.) 273*, 245–349 (1978).

Fonagy, A., M. G. Ord, & L. A. Stocken, *Biochem. J. 162*, 171–181 (1977).

Ford, R. J., S. R. Mehta, D. Franzini, R. Montagne, L. B. Lachman, & A. L. Maizel, *Nature (Lond.) 294*, 261–263 (1981).

Forsdyke, D. R., *Biochem. J. 107*, 197–206 (1968a).

Forsdyke, D. J., *Biochem. J. 108*, 297–302 (1968b).

Fowler, E., R. Farb, & S. El-Saidy, *Nucleic Acids Res. 10*, 735–748 (1982).

Fox, C. F., P. S. Linsley, & M. Wrann, *Fed. Proc. Fed. Am. Soc. Exp. Biol. 41*, 2988–2995 (1982).

Fox, I. H., & W. N. Kelley, *Annu. Rev. Biochem. 47*, 655–686 (1978).

Frankin, S. G., & A. Zweidler, *Nature (Lond.) 266*, 273–275 (1977).

Freedman, R. B., *FEBS Lett. 97*, 201–210 (1979).

French, V., P. J. Bryant, & S. V. Bryant, *Science 193*, 969–981 (1976).

Friedenstein, A. Y., *Clin, Orthop. 59*, 21–37 (1968).

Friedman, D. L., T. H. Clows, S. J. Pilkis, & G. E. Pine, *Exp. Cell Res. 135*, 283–290 (1981).

Fritsch, E. F., R. M. Lawn, & T. Maniatis, *Cell 19*, 959–972 (1980).

Füchtbauer, A., B. M. Jockusch, H. Maruta, M. W. Kiliman, & G. Isenberg, *Nature (Lond.) 304*, 361–364 (1983).

Gabbiani, G., C. Chaponnier, & I. Huttner, *J. Cell Biol. 76*, 561–568 (1978).

Gabrielli, F., R. Hancock, & A. J. Faber, *Eur. J. Biochem. 120*, 363–369 (1981).

Gabritchevsky, G., *Ann. Inst. Pasteur (Paris) 4*, 346 (1890).

Galli, G., H. Hofstetter, & M. L. Birnstiel, *Nature (Lond.) 294*, 626–631 (1981).

Gallwitz, D., & G. C. Mueller, *J. Biol. Chem. 244*, 5947–5952 (1969).

Garcia-Bellido, A., *Dev. Biol. 14*, 278–306 (1966).

Garcia-Bellido, A., & P. Ripoll, in *Results & Problems in Cell Differentiation 9*, 119–156 (1978).

Gartner, S., & H. S. Kaplan, *Proc. Natl. Acad. Sci. 77*, 4756–4759 (1980).

Gavryck, W. A., R. D. Moore, & R. C. Thompson, *J. Physiol. 252*, 43–58 (1975).

Geahlen, R. L., D. F. Carmichael, E. Hashimoto, & E. G. Krebs, *Adv. Enzyme Regul. 20*, 195–209 (1982).

Gehring, W. J., & R. Nöthiger, in *Developmental Systems: Insects*, S. J. Councie & C. H. Waddington, eds., Academic Press, New York, pp. 211–290 (1973).

Georgieva, E. I., I. G. Pashev, & R. G. Tsanev, *Arch. Biochem. Biophys. 216*, 88–92 (1982).

Gerber-Huber, S., F. E. B. May, B. R. Westley, B. K. Felber, H. A. Hosbach, A. C. Andres, & G. U. Ryffel, *Cell 33*, 43–51 (1983).

Gerhart, J. C., in *Biological Regulation and Development 2*, R. F. Goldberger, ed., Plenum, New York, pp. 133–316 (1980).

Gierer, A., *Curr. Top. Dev. Biol. 11*, 17–59 (1977).

Gierer, A., *Prog. Biophys. Mol. Biol. 37*, 1–47 (1981).

Gierer, A., & H. Meinhardt, *Kybernetik, 12*, 30–39 (1972).

Giri, C. P., M. H. P. West, & M. E. Smulson, *Biochemistry 17*, 3501–3504 (1978).

Gjerser, R., C. Gorka, S. Hasthorpe, J. T. Lawrence, & H. Eisen, *Proc. Natl. Acad. Sci. 79*, 2333–2337 (1982).

Goetzl, E. J., & K. F. Austen, *Proc. Natl. Acad. Sci. 72*, 4123–4127 (1975).

Goldberg, A. L., & A. C. St. John, *Annu. Rev. Biochem. 45*, 747–803 (1976).

Goldberg, N. D., M. K. Haddox, E. Durham, C. Lopez, & J. W. Hadden, in *Control of Proliferation in Animal Cells*, B. Clarkson & R. Baserga, eds., Cold Spring Harbor Laboratory, Cold Spring Harbor, New York, pp. 3024–3027 (1974).

Goldfine, I. D., *Biochim. Biophys. Acta. 650*, 53–67 (1981).

Goldknopf, I. L., & H. Busch, *Proc. Natl. Acad. Sci. 74*, 864–868 (1977).

Goldschneider, I., *Curr. Top. Dev. Biol. 14*, 33–57 (1980).

Goldsmith, M. E., *Nucleic Acids Res. 9*, 6471–6485 (1981).

Goldstein, J. L., & M. S. Brown, *Annu. Rev. Biochem. 46*, 897–930 (1977).

Goldstein, J. L., T. L. K. Low, G. B. Thurman, M. M. Zatz, N. Hall, J. Chen, S.-K. Hu, P. B. Naylor, & J. E. McClure, *Recent Prog. Horm. Res. 37*, 369–412 (1981).

Goodwin, B. C., *Analytical Physiology of Cells and Developing Organisms*, Academic Press, New York (1976).

Goodwin, B. C., & M. H. Cohen, *J. Theor. Biol. 25*, 49–107 (1969).

Goodwin, G. M., J. M. Walker, & E. W. Johns, in *The Cell Nucleus*, H. Busch, ed., Academic Press, New York, pp. 181–219 (1978).

Gordon, J., P. J. Nielsen, K. L. Manchester, H. Tawbin, L.-J. de Asua, & G. Thomas, *Curr. Top. Cell Regul. 21*, 89–99 (1982).

Gospodarowicz, D., & A. L. Mescher, *Annu. N.Y. Acad. Sci. 339*, 151–174 (1980).

Gospodarowicz, D., & J. S. Moran, *Annu. Rev. Biochem. 45*, 531–558 (1976).

Goss, R. J., *Principles of Regeneration*, Academic Press, New York, 1969.

Gottesfeld, J. M., & L. S. Bloomer, *Cell 21*, 751–760 (1980).

Gottesfeld, J. M., W. T. Garrard, G. Bagi, R. F. Wilson, & J. Bonner, *Proc. Natl. Acad. Sci. 71*, 2193–2199 (1974).

Gottesfeld, J. M., & G. A. Partington, *Cell 12*, 953–962 (1977).

Govindom, M. V., E. Spiess, & J. Majors, *Proc. Natl. Acad. Sci. 79*, 5157–5161 (1982).

Goyns, M. H., *J. Theor. Biol. 97*, 577–589 (1982).

Grady, L. J., W. P. Campbell, & A. B. North, *Eur. J. Biochem. 115*, 241–245 (1981).

Graham, J. M., M. C. B. Sumner, D. H. Curtis, & C. A. Pasternak, *Nature (Lond.) 246*, 291–295 (1973).

Grahame-Smith, D. G., R. W. Butcher, R. L. Ney, & E. W. Sutherland, *J. Biol. Chem. 242*, 5535–5541 (1967).

Green, H., *Cell 15*, 801–811 (1978).

Greenburg, G., & D. Gospodarowicz, *Exp. Cell Res. 140*, 1–14 (1982).

Gremigni, V., & C. Miceli, *Wilhelm Roux's Arch. Dev. Biol. 188*, 107–113 (1980).

Grinde, B., & R. Johnson, *Biochem. J. 202*, 191–196 (1982).

Grisham, J. W., R. L. Tillman, A. E. H. Nagel, & J. Compagno, in *Liver Regeneration after Experimental Injury*, R. Lesch & W. Reutter, eds., Stratton, New York, pp. 6–23 (1975).

Grosveld, G. C., E. deBoer, C. F. Shewmaker, & R. A. Flavell, *Nature (Lond.) 295*, 120–126 (1982).

Groudine, M., & H. Weintraub, *Cell 30*, 131–149 (1982).

Groudine, M., & H. Weintraub, *Proc. Natl. Acad. Sci. 77*, 5351–5354 (1980).

Guasch, M. D., M. Plana, & E. Itarte, *Biochem. Biophys. Res. Commun. 107*, 82–88 (1982).

Guidotti, G. G., A. F. Borghetti, & G. C. Gazola, *Biochim. Biophys. Acta 515*, 329–366 (1978).

Gulati, A. K., A. A. Zalewski, & A. H. Reddi, *Dev. Biol. 96*, 355–365 (1983).

Gunn, J. M., J. B. Bodner, S. E. Knowles, & F. J. Ballard, *Biochem. J. 210*, 251–258 (1983).

Gurdon, J. B., *The Control of Gene Expression in Animal Development*, Clarendon Press, Oxford (1974).

Gurdon, J. B., C. Dingwall, R. A. Laskey, & L. J. Kurn, *Nature (Lond.) 299*, 652–653 (1982).

Gurley, L. R., R. A. Walters, & R. A. Tobey, *Arch. Biochem. Biophys. 148*, 633–641 (1972).

Gurley, L., R. Walters, & R. Tobey, *Biochem. Biophys. Res. Commun. 50*, 744–750 (1973).

Haddox, M. K., & D. H. Russell, *Proc. Natl. Acad. Sci. 78*, 1712–1716 (1981).

Hagenbüchle, O., M. Tosi, U. Schibler, R. Bovey, P. K. Wellauer, & R. A. Young, *Nature (Lond.) 289*, 643–646 (1981).

Halkerston, I. D. K., *Adv. Cyclic Nucleotide Res. 6*, 99–136 (1975).

Halleck, M. D., & L. R. Gurley, *Exp. Cell Res. 138*, 271–285 (1982).

Ham, G. H., & M. J. Veomett, *Mechanisms of Development*, Mosby, St. Louis, Missouri (1980).

Hämmerling, J., *Annu. Rev. Plant Physiol. 14*, 65–92 (1963).

Hämmerling, J., *Int. Rev. Cytol. 2*, 475–498 (1953).

Hämmerling, U., *Progress in Allergy 28*, 40–65 (1981).

Handlogten, M. E., & M. Kilberg, *Biochem. Biophys. Res. Commun. 108*, 1113–1119 (1982).

Harkness, R. D., *Br. Med. Bull. 13*, 87–93 (1957).

Harris, H., *Cell Fusion*, Clarendon Press, Oxford (1970).

Harris, H., *Nucleus and Cytoplasm*, 3rd ed., Clarendon Press, Oxford (1974).

Hartwell, L. H., *J. Cell Biol. 77*, 627–637 (1978).

Hassid, A., *Am. J. Physiol C. 12*, 205–211 (1982).

Hathaway, G. M., T. S. Lundak, S. M. Tahara, & J. A. Traugh, *Methods Enzymol. 60*, 495–511 (1979).

Hathaway, G. M., & J. A. Traugh, *Curr. Top. Cell. Regul. 21*, 101–127 (1982).

Hay, E. D., *Regeneration*, Holt, Rinehart & Winston, New York (1966).

Hayaishi, O., & K. Ueda, *Annu. Rev. Biochem. 46*, 95–116 (1977).

Hayashi, K., M. Tanaki, T. Shimada, M. Miwa, & T. Sugimura, *Biochem. Biophys. Res. Commun. 112*, 102–107 (1983).

Heath, J. P., *Nature (Lond.) 302*, 532–534 (1983).

Heath, J. P., & G. A. Dunn, *J. Cell Sci. 29*, 197–212 (1978).

Heimfeld, S., & H. R. Bode, *J. Cell Sci. 52*, 85–98 (1981).

Heller, J. S., W. F. Fong, & E. S. Canellakis, *Proc. Natl. Acad. Sci. 73*, 1858–1862 (1976).

Henderson, D., H. Eibl, & K. Weber, *J. Mol. Biol. 132*, 193–218 (1979).

Hentschel, C. C., & M. L. Birnstiel, *Cell 25*, 301–313 (1981).

Herbert, E., & M. Uhler, *Cell 30*, 1–2 (1982).

Herbomel, P., S. Saragosti, D. Blangy, & M. Yaniv, *Cell 25*, 651–658 (1981).

Hereford, L., S. Bromley, & M. A. Osley, *Cell 30*, 305–310 (1982).

Hereford, L. M., M. A. Osley, J. L. Ludwig, & C. S. McLaughlin, *Cell 24*, 367–375 (1981).

Hershko, A., P. Mamont, R. Shields, & G. M. Tomkins, *Nature (Lond.) New Biol. 232*, 206–211 (1971).

Hertzberg, E. L., T. S. Lawrence, & N. B. Gilula, *Annu. Rev. Physiol. 43*, 479–491 (1981).

Hesketh, T. R., G. A. Smith, M. D. Houslay, G. B. Warren, & J. C. Metcalfe, *Nature (Lond.) 267*, 490–494 (1977).

Hewitt, J. A., *J. Theor. Biol. 80*, 115–127 (1979).

Higgins, G. M., & R. M. Anderson, *Arch. Path. 12*, 186–202 (1931).

Hilz, H., *Hoppe-Seyler's Z. Physiol. Chem. 362*, 1415–1425 (1981).

Hilz, H., P. Adamietz, R. Bredehorst, & K. Wielckens, *Adv. Enzyme Regul. 17*, 195–211 (1979).

Hilz, H., & P. Stone, *Rev. Physiol. Biochem. Pharmacol. 76*, 1–58 (1976).

Hirato, F., & J. Axelrod, *Science 209*, 1082–1090 (1980).

Hirsch, H. A., *J. Theor. Biol. 100*, 399–410 (1983).

Hofstetter, H., A. Kressman, & M. Birnstiel, *Cell 24*, 573–585 (1981).

Hohmann, P., R. A. Tobey, & L. R. Gurley, *J. Biol. Chem. 251*, 3685–3692 (1976).

Hokin, L. E., & M. R. Hokin, *Biochim. Biophys. Acta 18*, 102–110 (1955).

Holder, N., *Br. Med. Bull. 37*, 227–232 (1981).

Holder, N., S. V. Bryant, & P. W. Tank, *J. Exp. Zool. 208*, 303–310 (1979).

Holder, N., P. W. Tank, & S. V. Bryant, *Dev. Biol. 74*, 302–314 (1980).

Holley, R. W., *Nature (Lond.) 258*, 487–490 (1975).

Holley, R. W., P. Bohlen, R. Fava, J. H. Baldwin, G. Kleeman, & R. Armour, *Proc. Natl. Acad. Sci. 77*, 5989–5992 (1980).

Holliday, R., & J. E. Pugh, *Science 187*, 226–232 (1975).

Holmes, B. E., in *CIBA Foundation Symposium on Ionizing Radiation and Cell Metabolism*, Churchill, London, pp. 225–236 (1956).

Höltä, E., & J. Jänne, *FEBS Lett. 23*, 117–121 (1972).

Holtzer, H., S. Dienstman, J. Biehl, & S. Holtzer, in *Extracellular Matrix Influences on Gene Expression*, H. C. Slavkin & R. C. Greulich, eds., Academic Press, New York, pp. 253–257 (1975).

Holzer, H., & P. C. Heinrich, *Annu. Rev. Biochem. 49*, 63–91 (1980).

Hopgood, M. F., M. G. Clark, & F. J. Ballard, *Biochem. J. 196*, 33–40 (1981).

Hopkins, H. A., R. J. Bonney, P. R. Walker, J. D. Yager, & V. R. Potter, *Adv. Enzyme Regul. 11*, 169–191 (1973).

Houslay, M. D., & K. K. Stanley, *Dynamics of Biological Membranes*, Wiley, New York (1982).

Howard, A., & S. R. Pelc, *Exp. Cell Res. 2*, 178–187 (1951).

Hubbard, S. C., & R. J. Ivatt, *Annu. Rev. Biochem. 50*, 553–583 (1981).

Hunting, D., & J. F. Henderson, *Can. J. Biochem. 59*, 830–837 (1981).

Huttner, W. B. *Nature (Lond.) 299*, 273–276 (1982).

Huvasa, T., M. Segawa, N. Yamaguchi, & K. Oda, *J. Biochem. (Tokyo) 89*, 1375–1389 (1981).

Hyman, B. T., L. L. Stoll, & A. A. Spector, *Biochim. Biophys. Acta 713*, 375–385 (1982).

Hynes, R. O., *Biochim. Biophys. Acta 458*, 73–107 (1976).

Hynes, R. O., & K. M. Yamada, *J. Cell Biol. 95*, 369–377 (1982).

Igo-Kemenes, T., W. Horz, & H. G. Zachau, *Annu. Rev. Biochem. 51*, 89–121 (1982).

Ingebritsen, T. S., & P. Cohen, *Eur. J. Biochem. 132*, 255–261 (1983).

Ito, K., & H. Uchino, *J. Biol. Chem. 248*, 4782–4785 (1973).

Iversen, O. H., K. S. Bhangoo, & K. Hansen, *Virchows Arch. B. Cell Path. 16*, 157–179 (1974).

Iversen, O. H., R. Bjerknes, & F. Devik, *Cell Tissue Kinet. 1*, 351–367 (1968).

Jackson, D. A., S. J. McCready, & P. R. Cook, *Nature (Lond.) 292*, 552–555 (1981).

Jackson, R. C., M. S. Lui, T. J. Boritzki, H. P. Morris, & G. Weber, *Cancer Res. 40*, 1286–1291 (1980).

Jackson, V., & R. Chalkley, *Cell 23*, 121–134 (1981).

Jacobson, L. O., E. Goldwasser, W. Fried, & L. Plzak, *Nature (Lond.) 179*, 633–634 (1957).

Jaeger, A., & C. C. Kuenzle, *EMBO J. 1*, 811–816 (1982).

Jagus, R., & J. E. Kay, *Eur. J. Biochem. 100*, 503–510 (1979).

Jähner, D., H. Stuhlmann, C. L. Stewart, K. Harbers, J. Löhler, I. Simon, & R. Jaenisch, *Nature (Lond.) 298*, 623–628 (1982).

Jeejeebhoy, K. N., *Br. Med. Bull. 37*, 11–17 (1981).

Jelinek, W. R., & C. W. Schmid, *Annu. Rev. Biochem. 51*, 813–844 (1982).

Jennings, M. A., & H. W. Florey, in *Pathology*, 4th ed., H. W. Florey, ed., Lloyd-Luke, London, pp. 480–548 (1970).

Jensenius, J. C., & A. F. Williams, *Nature (Lond.) 300*, 583–588 (1982).

John, P. C. L., C. A. Lambe, R. McGookin, & B. Orr, in *The Cell Cycle*, P. C. L. John, ed., Cambridge University Press, New York, pp. 185–221 (1981).

Johnson, R. T., & H. Harris, *J. Cell Sci. 5*, 603–624 (1969).

Johnson, R. T., & P. N. Rao, *Nature (Lond.) 226*, 717–722 (1970).

Johnston, R. N., & N. K. Wessells, *Curr. Top. in Develop. Biol. 16*, 165–206 (1980).

Jonas, H. A., R. C. Baxter, & L. C. Harrison, *Biochem. Biophys. Res. Commun. 109*, 463–470 (1982).

Josse, J., A. D. Kaiser, & A. Kornberg, *J. Biol. Chem. 236*, 864–875 (1961).

Jourdian, G. W., G. G. Sahagian, & J. Distier, *Biochem. Soc. Trans. 9*, 510–512. (1981).

Kahn, C. R., K. L. Baird, J. S. Flier, C. Grunfeld, J. T. Harmon, L. C. Harrison, F. A. Karlsson, M. Kasuga, G. L. King, W. C. Lang, J. M. Podskalny, & E. V. Van Obberghen, *Rec. Prog. Horm. Res. 37*, 477–533 (1981).

Kanatani, H. J., *Fac. Sc. Univ. Tokyo 4*, 254–270 (1958).

Kaplan, J. G., *Ann. N.Y. Acad. Sci. 339*, 251–252 (1980).

Kaplan, J. G., *Annu. Rev. Physiol. 40*, 19–41 (1978).

Kaplan, J. G., & T. Owens, *Ann. N.Y. Acad. Sci. 339*, 191–200 (1980).

Kaser, M. R., M. G. Ord, & L. A. Stocken, *Biochem. Intern. 1*, 148–154 (1980).

Kasuga, M., Y. Zick, D. L. Blith, F. A. Karlsson, H. U. Häring, & C. R. Kahn, *J. Biol. Chem. 257*, 9891–9894 (1982).

Katz, D. H., *Adv. Immunol. 28*, 137–207 (1979).

Kauffman, S. A., R. M. Shymko, & K. Trabert, *Science 199*, 259–270 (1978).

Kay, J. E., *Biochem. Soc. Trans. 4*, 1120–1122 (1976).

Kecskemethy, N., & K. P. Schafer, *Eur. J. Biochem. 126*, 573–582 (1982).

Kedes, L. H., *Annu. Rev. Biochem. 48*, 837–870 (1979).

Kefalides, N. A., *Biology and Chemistry of Basement Membranes*, N. A. Kefalides, ed., Academic Press, Academic Press, New York, pp. 215–228 (1978).

Keller, H. U., & E. Sorkin, *Int. Arch. Allergy and Appl. Immunol. 31*, 505–517 (1967).

Kepple, F., B. Allet, & H. Eisen, *Proc. Natl. Acad. Sci. 74*, 653–656 (1977).

Khym, J. X., M. H. Jones, W. H. Lee, J. D. Regan, & E. Volkin, *J. Biol. Chem. 253*, 8741–8746 (1979).

Kiefer, H., & R. Schulze, *Biosci. Rep. 2*, 583–588 (1982).

Kivilaakso, E., & T. Rytöma, *Cell Tissue Kinet. 4*, 1–9 (1971).

Klee, C. B., T. H. Crouch, & P. G. Richman, *Annu. Rev. Biochem. 49*, 489–515 (1980).

Kleitzen, R. F., M. W. Pariza, J. E. Becker, & V. R. Potter, *J. Biol. Chem. 251*, 3014–3020 (1976).

Klenow, S., *Biochim. Biophys. Acta 699*, 49–52 (1982).

Knutton, S., *Exp. Cell Res. 102*, 109–116 (1976).

Koch, G. L. E., *Cell Adhesion and Mobility*, A. S. G. Curtis, & J. D. Pitts, eds., Cambridge University Press, Cambridge, England, pp. 425–444 (1980).

Koch, K., & H. L. Leffert, *J. Cell Biol. 62*, 780–791 (1974).

Kornberg, R., *Nature (Lond.) 292*, 579–580 (1981).

Kostka, G., & A. Schweiger, *Biochim. Biophys. Acta 696*, 139–144 (1982).

Kreis, T. E., B. Geiger, E. Schmid, J. L. Jorcano, & W. W. Franke, *Cell 32*, 1125–1137 (1983).

Kretsinger, R. H., *Annu. Rev. Biochem. 45*, 239–266 (1976).

Kuehl, L. R., T. Lyness, G. H. Dixon, & B. Levy-Wilson, *J. Biol. Chem. 255*, 1090–1095 (1980).

Kun, E., T. Minaga, E. Kirstein, G. Jackowski, J. McLick, L. Peller, S. M. Oredsson, L. Martin, N. Pattabiraman, & C. Milo, *Adv. Enzyme Regul. 21*, 177–199 (1983).

La Brecque, D. R., *Am. J. Physiol. G. 5*, 289–295 (1982).

La Brecque, D. R., & R. B. Howard, *Methods in Cell Biol. 14*, 327–340 (1976).

Lackie, J. M., in *Cell Behavior*, R. Bellairs, A. Curtis, & G. Dunn, eds., Cambridge University Press, Cambridge, England, pp. 319–348 (1982).

Laine, B., P. Sautiere, & G. Biserte, *Biochemistry 15*, 1640–1645 (1976).

Lajtha, L. G., *J. Cell. Comp. Physiol. 62*, 143–145 (1963).

Laks, M. S., & R. A. Jungmann, *Biochem. Biophys. Res. Commun. 96*, 697–703 (1980).

Lane, E. B., S. L. Goodman, & L. K. Trejdosiewicz, *EMBO J. 1*, 1365–1372 (1982).

Laperche, Y., K. R. Lynch, K. P. Dolan, & P. Feigelson, *Cell 32*, 453–460 (1983).

Larner, J., K. Cheng, C. Schwartz, K. Kikuchi, S. Tamura, S. Creacy, R. Dubler, G. Galasko, C. Pullin, & M. Katz, *Rec. Prog. Horm. Res. 38*, 511–552 (1982).

Larsen, A., & H. Weintraub, *Cell 29*, 609–622 (1982).

Laz, M., & R. E. Sicard, *Wilhelm Roux's Archiv. of Dev. Biol. 191*, 163–168 (1982).

Lazarides, E., *Annu. Rev. Biochem. 51*, 219–250 (1982).

Leader, D. P., in *Molecular Aspects of Cellular Regulation 1*, P. Cohen, ed., Elsevier North-Holland, New York, pp. 203–233 (1980).

Lebkowski, J. S., & U. K. Laemmli, *J. Mol. Biol. 156*, 325–344 (1982).

Leder, P., *Sci. Am. 246*, 72–83 (1982).

Lee, F., R. Mulligan, P. Berg, & G. Ringold, *Nature (Lond.) 294*, 228–232 (1981).

Lee, K. L, S. C. Sun, & O. N. Miller, *Archiv. Biochem. Biophys. 125*, 751–757 (1968).

Leffak, I. M., *Nucleic Acids Res. 11*, 2717–2731 (1983).

Leffak, I. M., R. Grainger, & H. Weintraub, *Cell 12*, 837–845 (1977).

Leffert, H. L., *J. Cell Biol. 62*, 767–779 (1974a).

Leffert, H. L., *J. Cell Biol. 62*, 792–801 (1974b).

Leffert, H. L., & K. S. Koch, *Ann. N.Y. Acad. Sci. 339*, 201–215 (1980).

Lenfont, M., M. P. de Garilha, E. Garcia-Giralt, & C. Tempête, *Biochim Biophys. Acta 451*, 106–117 (1976).

Lennox, R. W., & L. H. Cohen, *J. Biol. Chen. 258*, 262–268 (1983).

Leof, E. B., N. E. Olashaw, W. J. Pledger, & E. J. O'Keefe, *Biochem. Biophys. Res. Commun. 109*, 83–91 (1982).

Leong, G. F., J. W Gresham, B. V. Hole, & M. L. Albright, *Cancer Res. 24*, 1496–1501 (1964).

Lesch, R., W. Bachmann, & W. Reutter, *Cell Tissue Kinet. 6*, 315–323 (1973).

Levy, R., S. Levy, S. A. Rosenberg, & R. T. Simpson, *Biochemistry 12*, 224–228 (1973).

Lewis, J., J. M. W. Slack, & L. Wolpert, *J. Theor. Biol. 65*, 579–590 (1977).

Lichtler, A. C., F. Sierra, S. Clark, J. R. E. Wells, J. L. Stein, & G. S. Stein, *Nature (Lond.) 298*, 195–198 (1982).

Lienhard, G. E., *Trends in Biochem. Sci. 8*, 125–127 (1983).

Ling, N. R., *Lymphocyte Stimulation*, Wiley, New York, (1968).

Ling, N. R., & J. E. Kay, *Lymphocyte Stimulation*, 2d ed., Elsevier North-Holland, New York (1975).

Little, J. W., & D. W. Mount, *Cell 29*, 11–22 (1982).

Liversage, R. A., & B. J. Brandes, *Wilhelm Roux's Arch. Dev. Biol. 191*, 149–159 (1982).

Lloyd, D., R. K. Poole, & S. W. Edwards, *The Cell Division Cycle*, Academic Press, New York (1982).

Loewenstein, W. R., *Biochim. Biophys. Acta 560*, 1–65 (1979).

Lohr, D., *Biochemistry 22*, 927–934 (1983).

Long, B. H., C. Y. Huang, & A. O. Pogo, *Cell 18*, 1079–1090 (1979).

Long, E. O., & I. B. Dawid, *Annu. Rev. Biochem. 49*, 727–764 (1980).

Lord, B. I., N. G. Testa, & J. H. Hendry, *Blood 46*, 65–77 (1975).

Louis, C., & C. E. Sekeris, *Exp. Cell Res. 102*, 317–328 (1976).

Low, D. A., R. W. J. Scott, J. B. Baker, & D. D. Cunningham, *Nature (Lond.) 298*, 476–478 (1982).

Low, T. L. K., S. S. Wang, & A. L. Goldstein, *Biochemistry 22*, 733–740 (1983).

Lubin, M., F. Cahn, & B. A. Countermarsh, *J. Cell Physiol. 113*, 247–251 (1982).

Lund, E. J., *J. Exp. Zool. 24*, 1–34 (1917).

MacCabe, J. A., J. Errick, & J. W. Saunders, *Dev. Biol. 39*, 69–82 (1974).

MacCabe, J. A., & K. E. V. Richardson, *J. Embryol. Exp. Morphol. 67*, 1–12 (1982).

Macdonald, R. G., & J. A. Cidlowski, *Biochim. Biophys. Acta 678*, 18–26 (1981).

MacManus, J. P., B. M. Braceland, R. H. Rixon, J. F. Whitfield, & H. P. Morris, *FEBS Lett. 133*, 99–102 (1981).

MacManus, J. P., D. J. Franks, T. Youdale, & B. M. Braceland, *Biochem. Biophys. Res. Commun. 49*, 1201–1207 (1972).

MacManus, J. P., & J. F. Whitfield, *Exp. Cell Res. 58*, 188–191 (1969).

MacManus, J. P., J. F. Whitfield, & T. Youdale, *J. Cell Physiol. 77*, 103–116 (1969).

MacWilliams, H. K., *J. Theor. Biol. 99*, 681–703 (1982).

Maden, M., *J. Theor. Biol. 69*, 735-753 (1977).

Maden, M., *Nature (Lond.) 295*, 672–675 (1982).

Maden, M., & K. Mustafa, *J. Embryol. Exp. Morphol. 70*, 197–213 (1982).

Mandelstam, J., *Bacteriol. Rev. 24*, 289–309 (1960).

Mandelstam, J., K. McQuillen, & I. W. Dawes, *Biochemistry of Bacterial Growth*, 3d ed., Blackwell Scientific Publications, Boston (1982).

Marchmont, R. J., & M. D. Houslay, *Biochem. J. 195*, 653–660 (1981).

Marcu, K. B., & M. D. Cooper, *Nature (Lond.) 298*, 327–328 (1982).

Mardian, J. K. W., A. E. Patern, G. J. Bunick, & D. E. Olins, *Science 209*, 1534–1536 (1980).

Mariani, B. D., D. L. Slate, & R. T. Schimke, *Proc. Natl. Acad. Sci. 78*, 4985–4989 (1981).

Marks, D. B., W. K. Paik, & T. W. Borun, *J. Biol. Chem. 248*, 5660–5667 (1973).

Marks, P. A., & R. A. Rifkind, *Annu. Rev. Biochem. 47*, 419–448 (1978).

Marks, P. A., R. A. Rifkind, R. Gambari, E. Epner, Z. Chen, & J. Banks, *Curr. Top. Cell Regul. 21*, 189–203 (1982).

Marsden, M. P. F., & U. K. Laemmli, *Cell 17*, 849–858 (1979).

Marsh, W. H., & P. Fitzgerald, *Fed. Proc. Am. Soc. Exp. Biol. 32*, 2119–2125 (1973).

Martin, D., G. M. Tomkins, & D. Granner, *Proc. Natl. Acad. Sci. 62*, 248–255 (1969).

Martin-Perez, J., & G. Thomas, *Proc. Natl. Acad. Sci. 80*, 926–930 (1983).

Marx, J. L., *Science 212*, 653–655 (1981).

Mathis, D., P. Oudet, & P. Chambon, *Prog. Nucleic Acids Res. Mol. Biol. 24*, 1–55 (1980).

Matthews, H. R., in *Recently Discovered Systems of Enzyme Regulation by Reversible Phosphorylation*, P. Cohen, ed., Elsevier North-Holland, New York, pp. 235–254 (1980).

Mazia, D., in *The Cell, III*, J. Brachet & A. E. Mirsky, eds., Academic Press, New York, pp. 77–412 (1961).

McCulloch, E. A., & J. E. Till, *Am. J. Path. 65*, 601–619 (1971).

McGhee, J. D., & G. Felsenfeld, *Annu. Rev. Biochem. 49*, 1115–1156 (1980).

McGhee, J. D., & G. Felsenfeld, *Cell 32*, 1205–1215 (1983).

McGhee, J. D., D. C. Rau, E. Charney, & G. Felsenfeld, *Cell 22*, 87–96 (1980).

McGhee, J. D., W. I. Wool, M. Dolan, J. D. Engel, & G. Felsenfeld, *Cell 27*, 45–55 (1981).

McGinnis, W., A. W. Shermoen, J. Heems-Kerk, & S. K. Beckendorf, *Proc. Natl. Acad. Sci. 80*, 1063–1067 (1983).

McGowan, J., V. Atryzck, & N. Fausto, *Biochem. J. 180*, 25–35 (1979).

McGowan, J. A., & N. Fausto, *Biochem. J. 170*, 123–127 (1978).

McGowan, J. A., A. J. Strain, & N. L. R. Bucher, *J. Cell Physiol. 108*, 353–363 (1981).

McMahon, J. B., J. G. Farrell, & P. T. Iype, *Proc. Natl. Acad. Sci. 79*, 456–460 (1982).

Meedel, T. H., & E. M. Levine, *J. Cell Physiol. 94*, 229–242 (1978).

Meggio, F., A. D. Deana, A. M. Brunati, & L. A. Pinna, *FEBS Lett. 141*, 257–262 (1982).

Meggio, F., A. D. Deana, & L. A. Pinna, *Biochim. Biophys. Acta 662*, 1–7 (1981).

Melchner, von H., & E. J. Lieschka, *Blook 57*, 906–912 (1981).

Melli, M., G. Spinelli, & E. Arnold, *Cell 12*, 167–174 (1977).

Merlie, J. P., M. E. Buckingham, & R. G. Whalen, *Curr. Top. Dev. Biol. 11*, 61–114 (1977).

Mészáros, G., T. Romhányi, T. F. Antoni, & A. Faragó, *FEBS Lett. 134*, 139–142 (1981).

Meyer, D. J., S. B. Yancey, & J. P. Revel, *J. Cell Biol. 91*, 505–523 (1981).

Michell, R. H., *Biochim. Biophys. Acta 415*, 81–147 (1975).

Millis, A. J. T., G. A. Forrest, & D. A. Pious, *Exp. Cell Res. 83*, 335–343 (1974).

Milstein, C., *Proc. Roy. Soc. London Ser. B 211*, 393–412 (1981).

Minakuchi, R., Y. Takai, B. Yu, & Y. Nishizuka, *J. Biochem, (Tokyo) 89*, 1651–1654 (1981).

Mirand, E. A., in *Regulation of Organ and Tissue Growth*, R. J. Goss, ed., Academic Press, New York, pp. 143–158 (1972).

Mitchison, J. M., *The Biology of the Cell Cycle*, Cambridge University Press, Cambridge, England (1971).

Mohun, T. J., R. Tilly, R. Mohun, & J. M. W. Slack, *Cell 22*, 9–15 (1980).

Moll, R., W. W. Franke, D. L. Schiller, B. Geiger, & R. Krepler, *Cell 31*, 11–24 (1982).

Moment, T. H., *Am. Nat. 87*, 139–153 (1953).

Moolenaar, W. H., C. L. Mummery, P. T. van der Saags, & S. W. de Laat, *Cell 23*, 789–798 (1981).

Moore, R. D., *Biochim. Biophys. Acta 737*, 1–49 (1983).

Morata, G., & S. Kerridge, *Nature (Lond.) 300*, 191–192 (1982).

Morgan, T. H., *J. Exp. Zool. 1*, 385–390 (1904).

Morgan, T. H., *Regeneration*, Macmillan, New York (1901).

Morgenegg, G., M. R. Celio, B. Malfoy, M. Leng, & C. C. Kuenzle, *Nature (Lond.) 303*, 540–543 (1983).

Morley, C. G. D., *Biochim. Biophys. Acta 362*, 480–492 (1974).

Morley, C. G. D., & H. S. Kingdon, *Biochim. Biophys. Acta 308*, 260–275 (1973).

Moss, M. L., in *Regulation of Organ and Tissue Growth*, R. J. Goss, ed., Academic Press, New York, pp. 127–142 (1972).

Moss, T., & M. L. Birnstiel, *Nucleic Acids Res. 6*, 3733–3743 (1979).

Mount, S. M., *Nucleic Acids Res. 10*, 459–472 (1982).

Mufson, R. A., E. G. Astrup, R. C. Simsiman, & R. K. Boutwell, *Proc. Natl. Acad. Sci. 74*, 657–661 (1977).

Müller, W. A., *Differentiation 22*, 141–150 (1982).

Munro, H. N., *Biochem. J. 113*, 1P (1969).

Munro, H. N., *N. Eng. J. Med. 300*, 41–42 (1979).

Murdoch. G. H., M. G. Rosenfeld & R. M. Evans, *Science 218*, 1315–1317 (1982).

Murray, A. B., W. Stecker, & S. Silz, *J. Cell Biol. 50*, 433–448 (1981).

Nabeshima, Y., & K. Ogata, *Eur. J. Biochem. 107*, 323–329 (1980).

Nadeau, P., D. R. Oliver, & R. Chalkey, *Biochemistry 17*, 4885–4893 (1978).

Nakanishi, S., T. Kita, S. Taii, H. Imura, & S. Numa, *Proc. Natl. Acad. Sci. 74*, 3283–3286 (1977).

Nakayasu, H., & K. Ueda, *Exp. Cell Res. 143*, 55–62 (1983).

Neufeld, N. D., M. Scott, & S. A. Kaplan, *Dev. Biol. 78*, 151–160 (1980).

Nicholas, R. H., C. A. Wright, P. N. Cockerill, J. A. Wyke, & G. H. Goodwin, *Nucleic Acids Res. 11*, 753–772 (1983).

Niedel, J. E., & P. Cuatrecasas, *Curr. Top. Cell Regul. 17*, 137–170 (1980).

Nienhuis, A. W., & E. J. Benz, *N. Eng. J. Med. 297*, 1318–1328, 1371–1381, 1430–1436 (1977).

Nilsen-Hamilton, M., R. T. Hamilton, W. R. Allen & S. Potter-Perigo, *Cell 31*, 237–242 (1982).

Nilsen-Hamilton, M., J. M. Shapiro, S. L. Massoglia, & R. T. Hamilton, *Cell 20*, 19–28 (1980).

Nishimura, J., & T. F. Deuel, *FEBS Lett. 156*, 130–134 (1983).

Nishio, A., S. Nakanishi, J. Doull, & E. M. Uyeki, *Biochem. Biophys. Res. Commun. 111*, 750–759 (1983).

Noonan, K. D., A. J. Levine, & M. M. Burger, *J. Cell Biol. 58*, 491–497 (1973).

Nordheim, A., M. L. Pardue, E. M. Lafer, A. Möller, B. D. Stollar, & A. Rich, *Nature (Lond.) 294*, 417–422 (1981).

Nordheim, A., P. Tesser, F. Azorin, Y. H. Kwon, A. Möller, & A. Rich, *Proc. Natl. Acad. Sci. 79*, 7729–7733 (1982).

Nurse, P. *Nature (Lond.) 286*, 9–10 (1980).

Nurse, P., & P. A. Fantes, in *The Cell Cycle*, P. C. L. John, ed., Cambridge University Press, Cambridge, England, pp. 85–98 (1981).

O'Dea, R. F., O. H. Viveros, J. Axelrod, S. Aswanikumar, E. Schiffmann, & B. A. Corcoran, *Nature (Lond.) 272*, 462–464 (1978).

Olsen, R. E., & J. W. Suttie, *Vitam. & Horm. 35*, 59–108 (1977).

Oppenheimer, C. L., J. E. Pessin, J. Massague, W. Gitomer, & M. P. Czech, *J. Biol. Chem. 258*, 4824–4830 (1983).

Ord, M. G., & L. A. Stocken, *Biochem. J. 103*, 5–6P (1967).

Ord, M. G., & L. A. Stocken, *Biochem. J. 129*, 175–181 (1972a).

Ord, M. G., & L. A. Stocken, *Biochem. J. 132*, 47–54 (1973a).

Ord, M. G., & L. A. Stocken, *Biochem. J. 136*, 571–577 (1973b).

Ord, M. G., & L. A. Stocken, *Biochem. J. 178*, 173–185 (1979).

Ord, M. G., & L. A. Stocken, *Biochem. Soc. Trans. 8*, 759–766 (1980).

Ord, M. G., & L. A. Stocken, *Biochem. Soc. Trans. 9*, 242–243 (1981b).

Ord, M. G., & L. A. Stocken, *Biosci. Rep. 1*, 217–222 (1981a).

Ord, M. G., & L. A. Stocken, in *Cell Biology and Medicine* E. E. Bittar, ed., Wiley, New York, pp. 151–195 (1972b).

Ord, M. G., & L. A. Stocken, in *Liver Regeneration after Experimental Injury*, R. Lesch & W. Reutter, eds., Stratton, New York, pp. 152–155 (1975a).

Ord, M. G., & L. A. Stocken, *Proc. FEBS Meet. 9th. 34*, 173–175 (1975b).

Ord, M. G., & L. A. Stocken, *Proc. Roy. Soc. Edin. Ser. B 70*, 117–124 (1968).

Ord, M. G., & L. A. Stocken, & S. Thrower, *Subcell Biochem. 4*, 147–156 (1975).

Osley, M. A., & L. M. Hereford, *Cell 24*, 377–384 (1981).

Osley, M. A., & L. M. Hereford, *Proc. Natl. Acad. Sci. 79*, 7689–7693 (1982).

Otteskog, P., T. Ege, & K. G. Sundquist, *Exp. Cell Res. 136*, 203–213 (1981).

Owen, J. J. T., & E. J. Jenkinson, *Adv. Immunol. 29*, 1–34 (1981).

Owen, M., *Int. Rev. Cytol. 28*, 213–238 (1970).

Paik, W. K., & S. Kim, *Protein Methylation*, Wiley-Interscience, New York (1979).

Pain, V. M., & M. J. Clemens, *Biochemistry 22*, 726–733 (1983).

Panyim, S., & R. Chalkley, *Arch. Biochem. Biophys. 130*, 337–346 (1969).

Panyim, S., & R. Chalkley, *Biochemistry 8*, 3972–3979 (1969).

Pardee, A. B., R. Dubrow, J. L. Hamlin, & R. F. Kletzien, *Annu. Rev. Biochem. 47*, 715–750 (1978).

Pardoll, D. M., B. Vogelstein, & D. S. Coffey, *Cell 19*, 527–536 (1980).

Pastan, I. H., G. S. Johnson, & W. B. Anderson, *Annu. Rev. Biochem. 44*, 491–522 (1975).

Pasternak, C. A., M. C. B. Sumner, & R. C. L. S. Collin, *Cell Cycle Controls*, G. M. Padilla, I. L. Cameron, & A. Zimmerman, eds., Academic Press, New York, pp. 117–124 (1974).

Paul, J., *Biosci. Rep. 2*, 63–76 (1982).

Paulson, J. R., & U. K. Laemmli, *Cell 12*, 817–828 (1977).

Payvar, F., O. Wrange, J. Carlstedt-Duke, S. Okrer, J. A. Gustafsson, & K. R. Yamamoto, *Proc. Natl. Acad. Sci. 78*, 6628–6632 (1981).

Pearse, B. M. F., & M. S. Bretscher, *Annu. Rev. Biochem. 50*, 85–101 (1981).

Perry, M., & R. Chalkley, *J. Biol. Chem. 256*, 3313–3318 (1981).

Peters, J. H., *Methods in Cell Biol. 9*, 1–11 (1975).

Peterson, D. R., & F. A. Carone, *Anat. Rec. 193*, 383–388 (1979).

Petruzelli, L. M., S. Ganguly, C. J. Smith, M. H. Cobb, C. S. Rubin, & O. M. Rosen, *Proc. Natl. Acad. Sci. 79*, 6792–6796 (1982).

Phillips, P. D., & V. J. Cristofalo, *Exp. Cell Res. 134*, 297–302 (1981).

Pickart, L., & M. M. Thaler, *Nature (Lond.) New Biol. 243*, 85–87 (1973).

Pierre, M., & J. E. Loeb, *Biochim. Biophys. Acta 700*, 221–228 (1982).

Pike, B. L., D. L. Vaux, I. Clark-Lewis, J. W. Schroeder, & G. J. V. Nossal, *Proc. Natl. Acad. Sci. 79*, 6350–6354 (1982).

Pike, M. C., & R. Snyderman, *Cell 28*, 107–114 (1982).

Plageman, P. G. W., & R. M. Wohlhueter, *Curr. Top. Membr. Trans. 14*, 225–330 (1980).

Pogo, B. G. T., & J. R. Katz, *Differentiation 2*, 119–124 (1974).

Poirier, G. G., G. de Murcia, J. Jongstra-Bilen, C. Niedergang, & P. Mandel, *Proc. Natl. Acad. Sci. 79*, 3423–3427 (1982).

Polunovsky, V. A., N. A. Setkov, & O. I. Epifanova, *Exp. Cell Res. 146*, 377–383 (1983).

Porter, K., D. Prescott, & J. Frye, *J. Cell Biol. 57*, 815–836 (1973).

Potten, C. S., & T. D. Allen, *Differentiation 3*, 161–165 (1975).

Prentice, D. A., S. C. Loechel, & P. A. Kitos, *Biochemistry 21*, 2412–2420 (1982).

Prentice, D. A., S. E. Taylor, M. Z. Newmark, & P. A. Kitos, *Biochem. Biophys. Res. Commun. 85*, 541–550 (1978).

Prescott, D. M., *Ann. N.Y. Acad. Sci. 397*, 101–109 (1982).

Prescott, D. M., *Reproduction of Eukaryotic Cells*, Academic Press, New York (1976).

Pressman, B. C., *Annu. Rev. Biochem. 45*, 501–530 (1976).

Pringle, J. R., & L. H. Hartwell, in *The Molecular Biology of the Yeast Saccharomyces*, J. N. Strathern, E. W. Jones, & J. R. Broad, eds., Cold Spring Harbor Laboratory, Cold Springs Harbor, New York, pp. 97–142 (1981).

Proudfoot, N. J., & G. G. Brownlee, *Nature (Lond.) 263*, 211–214 (1976).

Purnell, M., P. R. Stone, & W. J. D. Whish, *Biochem. Soc. Trans. 8*, 215–227 (1980).

Queensberry, P., & L. Levitt, *N. Eng. J. Med. 301*, 755–760, 819–823 (1979).

Quesney-Huneeus, V., H. A. Galick, M. D. Siperstein, S. K. Erickson, T. A. Spencer, & J. A. Nelson, *J. Biol. Chem. 258*, 378–385 (1983).

Quintans, J., & D. Kaplan, *Cell. Immunol. 40*, 236–242 (1978).

Rabes, H. M., H. V. Tuczek, & R. Wirsching, in *Liver Regeneration after Experimental Injury*, R. Lesch & W. Reutter, eds., Stratton, New York, pp. 35–52 (1975).

Racker, E., *Biosci. Rep. 3*, 507–516 (1983).

Radke, K., T. Gilmore, & G. S. Martin, *Cell 21*, 821–828 (1980).

Rao, M. V. N., *Int. Rev. Cytol. 67*, 291–315 (1980).

Rao, P. N., & R. T. Johnson, *Nature (Lond.)* 225, 159–164 (1970).

Rao, P. N., B. A. Wilson, & P. S. Sunkara, *Proc. Natl. Acad. Sci.* 75, 5043–5047 (1978).

Rasmussen, H., & D. Waisman, in *Biochemical Actions of Hormones, VIII*, G. Litwack, ed., Academic Press, New York, pp. 1–115 (1981).

Rath, N. C., & A. H. Reddi, *Nature (Lond.)* 278, 855–857 (1979).

Razin, A., & J. Friedman, *Prog. Nucleic Acids Res. Mol. Biol.* 25, 33–52 (1981).

Reddy, G. P., *Biochem. Biophys. Res. Commun.* 109, 908–915 (1982).

Reddy, G. P., & A. B. Pardee, *J. Biol. Chem.* 257, 12,526–12,531 (1982).

Reeves, R., & E. P. M. Candido, *Biochem. Biophys. Res. Commun.* 89, 571–579 (1979).

Reeves, R., & D. Chang, *J. Biol. Chem.* 258, 679–687 (1983).

Reichard, P., *Adv. Enzyme Regul.* 10, 3–16 (1972).

Repesh, L. A., T. J. Fitzgerald, & L. T. Furcht, *Differentiation* 22, 125–131 (1982).

Reynolds, C. H., *Biochem. J.* 202, 125–131 (1982).

Rickles, R., F. Marashi, F. Sierra, S. Clark, J. Wells, J. Stein, & G. Stein, *Proc. Natl. Acad. Sci.* 79, 749–753 (1982).

Riggs, A. D., *Cytogenet. Cell Genet.* 14, 9–25 (1975).

Rimmington, J. A., & D. H. Russell, *J. Cell Physiol.* 113, 252–256 (1982).

Ringer, D. P., D. E. Kizer, & R. L. King, *Biochim. Biophys. Acta* 656, 62–68 (1981).

Rixon, R. H., & J. F. Whitfield, *J. Cell Physiol.* 87, 147–156 (1976).

Rixon, R. H., & J. F. Whitfield, *J. Cell Physiol.* 113, 281–288 (1982).

Robbins, E., & T. W. Borun, *Proc. Natl. Acad. Sci.* 57, 409–416 (1967).

Robbins, S. L., M. Angell, & V. Kumar, *Basic Pathology*, 3d ed., Saunders, Philadelphia, Pennsylvania (1981).

Robertis, E. de., *Cell* 32, 1021–1025 (1983).

Roberts, J. L., C. L. C. Chen, J. H. Eberwine, M. R. Q. Evinger, C. Gee, E. Herbert, & B. S. Schachter, *Recent Prog. Horm. Res.* 38, 227–250 (1982).

Robertson, J. D., *J. Cell Biol.* 19, 201–221 (1963).

Robinson, S. I., B. D. Nelkin, & B. Vogelstein, *Cell* 28, 99–106 (1982).

Rocklin, R. E., K. Bendtzen, & D. Greineder, *Adv. Immunol.* 29, 55–136 (1980).

Roelants, G. E., in *B and T Cells in Immune Recognition*, F. Loor & G. E. Roelants, eds., Wiley, New York, pp. 103–125 (1977).

Roels, H., in *Tissue Repair and Regeneration*, L. E. Glynn, ed., Elsevier North-Holland, New York, pp. 243–283 (1981).

Romhányi, T., J. Seprödi, F. Mészáros, F. Antoni, & A. Faragó, *FEBS Lett.* 144, 223–225 (1982).

Rose, B., I. Simpson, & W. R. Loewenstein, *Nature (Lond.)* 267, 625–627 (1977).

Rose, F. C., & S. M. Rose, *Growth* 29, 361–393 (1965).

Rose, S. M., *Regeneration*, Appleton-Century-Crofts, New York (1970).

Rosen, O. M., R. Herbera, Y. Olowe, L. M. Petruzelli, & M. H. Cobb, *Proc. Natl. Acad. Sci.* 80, 3237–3240 (1983).

Rossow, P. W., V. G. H. Riddle, & A. B. Pardee, *Proc. Natl. Acad. Sci.* 76, 4446–4450 (1979).

Roth, R. A., & D. J. Cassell, *Science* 219, 299–301 (1983).

Rothenberg, P., L. Reuss, & L. Glaser, *Proc. Natl. Acad. Sci. 79*, 7783–7787 (1982).

Rothstein, H., *Int. Rev. Cytol. 78*, 127–232 (1982).

Rozengurt, E., *Curr. Top. Cell. Regul. 17*, 59–88 (1980).

Rozengurt, E., *J. Cell Physiol. 112*, 243–250 (1982).

Rozengurt, E., & S. Mendoza, *Ann. N.Y. Acad. Sci. 339*, 175–190 (1980).

Rubin, R. A., & H. S. Earp, *Science 219*, 60–63 (1983).

Rubin, R. A., E. J. O'Keefe, & H. S. Earp, *Proc. Natl. Acad. Sci. 79*, 776–780 (1982).

Russell, D. H., *Adv. Enzyme Regul. 21*, 201–222 (1983).

Russell, D. H., & B. G. M. Durie, *Polyamines as Biochemical Markers of Normal and Malignant Growth*, Rowen, New York (1978).

Russell, D. H., & M. K. Haddox, *Adv. Enzyme Regul. 17*, 61–87 (1979).

Russell, D. H., & P. J. Stambrook, *Proc. Natl. Acad. Sci. 72*, 1482–1486 (1975).

Russev, G., & R. Hancock, *Proc. Natl. Acad. Sci. 79*, 3143–3147 (1982).

Rytöma, T., & K. Kiviniemi, *Cell Tissue Kinet. 1*, 329–340, 341–350 (1968).

Sachsenmaier, W., in *The Cell Cycle*, P. C. L. Johns, ed., Cambridge University Press, Cambridge, England, pp. 139–160 (1981).

Salisbury, J. L., J. S. Condeelis, N. J. Maile, & P. Satir, *Nature (Lond.) 294*, 163–166 (1981).

Samuels, H. H., A. J. Perlman, B. M. Raaka, & F. Stanley, *Recent Prog. Horm. Res. 38*, 557–592 (1982).

Samuelsson, B., M. Goldyne, E. Granström, M. Hamberg, S. Hammarström, & C. Malmsten, *Annu. Rev. Biochem. 47*, 997–1029 (1978).

Sanger, J. W., & J. M. Sanger, *Cell & Tissue Res. 209*, 177–186 (1980).

Saragosti, S., G. Moyne, & M. Yaniv, *Cell 20*, 65–73 (1980).

Sargan, D. R., & P. H. W. Butterworth, *Nucleic Acids Res. 10*, 4641–4653, 4655–4669 (1982).

Sasi, R., P. E. Hüvös, & G. D. Fasman, *J. Biol. Chem. 257*, 11,448–11,454 (1982).

Schaller, C. H., *J. Embryol. Exp. Morphol. 29*, 27–38 (1973).

Scheer, U., *Cell 13*, 535–549 (1978).

Scher, W., B. M. Scher, & S. Waxman, *Biochem. Biophys. Res. Commun. 109*, 348–354 (1982).

Scherbaum, O., & E. Zeuthen, *Exp. Cell Res. 6*, 221–227 (1954).

Schibler, U., O. Hagenbüchle, P. K. Wellauer, & A. C. Pittet, *Cell 33*, 501–508 (1983).

Schiffman, E., B. A. Corcoran, & S. Aswanikumar, *Methods in Physiology and Clinical Applications*, J. I. Gallin & P. G. Quie, eds., Raven Press, New York, pp. 97–107 (1978).

Schiffman, Y., *Prog. Theor. Biol. 6*, 1–21 (1981).

Schilling, J. A., *Physiol. Rev. 48*, 374–423 (1968).

Schimmer, B. P., *Adv. Cyclic Nucleotide Res. 13*, 181–214 (1980).

Schlessinger, J., & B. Geiger, *Exp. Cell Res. 134*, 273–279 (1981).

Schlessinger, J., A. B. Schreiber, A. Levi, I. Lax, T. Liberman, & Y. Yarden, *Crit. Rev. Biochem. 14*, 93–111 (1983).

Schlosser, C. A., C. Steglich, J. R. de Wet, & I. E. Scheffler, *Proc. Natl. Acad. Sci. 78*, 1119–1123 (1981).

Schmidt, A. J., *Cellular Biology of Vertebrate Regeneration and Repair,* University of Chicago Press, Chicago (1968).

Schmidt, T., & H. C. Schaller, *Wilhelm Roux's Archiv. in Dev. Biol. 188,* 133–139 (1980).

Scholla, C. A., M. V. Tedeschi, & N. Fausto, *J. Biol. Chem. 255,* 2855–2860 (1980).

Schreiner, G. F., & E. R. Unanue, *Adv. Immunol. 24,* 37–165 (1976).

Schuldiner, S., & E. Rozengurt, *Proc. Natl. Acad. Sci. 79,* 7778–7782 (1982).

Schulster, D., & A. Levitski, *Cellular Receptors for Hormones and Neurotransmitters,* Wiley, New York (1980).

Schulster, D., & R. Schwyzer, in *Cellular Receptors for Hormones and Neurotransmitters,* D. Schulster & A. Levitski, eds., Wiley, New York, pp. 197–217 (1980).

Scornik, O. A., *J. Biol. Chem. 258,* 882–886 (1983).

Scornik, O. A., & V. Botbol, *J. Biol. Chem. 251,* 2891–2897 (1976).

Scott, R. E., B. M. Boman, D. E. Swartzendruber, M. A. Zschunke, & B. J. Hoerl, *Exp. Cell Res. 133,* 73–82 (1981).

Sealy, L., & R. Chalkley, *Archiv. Biochem. Biophys. 197,* 78–82 (1979).

Sefton, B. N., T. Hunter, E. H. Ball, & S. J. Singer, *Cell 24,* 165–174 (1981).

Seidman, M. M., A. J. Levine, & H. Weintraub, *Cell 18,* 439–449 (1979).

Sekas, G., W. G. Owen, & R. T. Cook, *Exp. Cell Res. 122,* 47–54 (1979).

Sengel, C., *J Embryol. Exp. Morphol. 8,* 468–476 (1960).

Severin, E. S., & M. V. Nesterova, *Adv. Enzyme Regul. 20,* 167–193 (1982).

Shafie, S. M., & R. Hilf, *Cancer Res. 41,* 826–829 (1981).

Sharma, R. K., *Prog. Nucleic Acids Res. Mol. Biol. 27,* 233–288 (1982).

Shepard, E. A., I. Phillips, J. Davis, J. L. Stein, & G. S. Stein, *FEBS Lett. 140,* 189–192 (1982).

Sheppard, J. R., & D. M. Prescott, *Exp. Cell Res. 72,* 293–296 (1972).

Shia, M. A., & P. F. Pilch, *Biochemistry 22,* 717–721 (1983).

Shields, R., *Nature (Lond.) 267,* 704–707 (1977).

Shields, R., R. F. Brooks, P. N. Riddle, D. F. Caparello, & D. Delia, *Cell 15,* 469–474 (1978).

Short, J., N. B. Armstrong, R. Zemes, & I. Lieberman, *Biochem. Biophys. Res. Commun. 50,* 430–437 (1973).

Short, J., R. F. Brown, A. Husakova, J. R. Gilbertson, R. Zemel, & I. Lieberman, *J. Biol. Chem. 247,* 1757–1766 (1972).

Shotwell, M. A., M. S. Kilberg, & D. L. Oxender, *Biochim. Biophys. Acta 737,* 267–284 (1983).

Siebenlist, U., J. V. Ravetch, S. Korsmeyer, T. Waldmann, & P. Leder, *Nature (Lond.) 294,* 631–635 (1981).

Siebert, G., M. G. Ord, & L. A. Stocken, *Biochem. J. 122,* 721–725 (1971).

Sigel, B., in *Regulation of Organ and Tissue Growth,* R. J. Goss, ed., Academic Press, New York, pp. 271–282 (1972).

Simpson, R. T., *Cell 13,* 691–699 (1978).

Singer, M., *Q. Rev. Biol. 27,* 169–200 (1952).

Singer, M. E., *Cell 28,* 433–434 (1982).

Skog, S., B Tribukait & G. Sunduis, *Exptl. Cell Res. 141,* 23–29 (1982).

Slack, J. M. W., *J. Embryol. Exp. Morphol. 70,* 241–260 (1982).

Slack, J. M. W., *J. Embryol. Exp. Morphol. 73,* 233–247 (1983).

Slack, J. M. W., *J. Theor. Biol. 82,* 105–140 (1980a).

Slack, J. M. W., *Nature (Lond.) 286,* 760 (1980b).

Slack, J. M. W., & S. Savage, *Nature (Lond.) 271,* 760–761 (1978).

Smart J. E., H. Opermann, A. P. Czernilofsky, A. F. Purchio, R. L. Erikson, & J. M. Bishop, *Proc. Natl. Acad. Sci. 78,* 6013–6017 (1981).

Smith, B. J., & E. W. Johns, *FEBS Lett. 110,* 25–29 (1980).

Smith, J. A., D. J. R. Lawrence, & P. S. Rowland, *Nature (Lond.) 293,* 648–650 (1981).

Smith, J. A., & L. Martin, *Proc. Natl. Acad. Sci. 70,* 1263–1267 (1973).

Smith, K. A., & F. W. Ruscetti, *Adv. Immunol. 31,* 137–175 (1981).

Smulson, M., *Trends in Biochem. Sci. 4,* 225–227 (1979).

Snyder, D. S., D. I. Beller, & E. R. Unanue, *Nature (Lond.) 299,* 163–165 (1982).

Söderhäll, S. S., A. Larsen, & K. L. Skoog, *Eur. J. Biochem. 33,* 36–39 (1973).

Solomon, F. *Cell 24,* 279–280 (1981).

Sons, W., H. J. Unsold, & R. Knippers, *Eur. J. Biochem. 65,* 263–269 (1976).

Sparfford, J. B., M. Ashburner, & E. Novitski, in *Genetics and Biology of Drosophila 1C,* M. Ashburner & T. R. F. Wright, eds., Academic Press, New York, pp. 955–1018 (1978).

Stadel, J. M., A. Delean, & R. J. Lefkowitz, *Adv. Enzymol. 53,* 1–43 (1982).

Stadler, J., A. Larson, J. D. Engel, M. Dolan, M. Groudine, & H. Weintraub, *Cell 20,* 451–460 (1980).

Stahl, H, & D. Gallwitz, *Eur. J. Biochem. 72,* 385–392 (1977).

Stein, G. S., W. Park, C. Thrall, R. Mans, & J. L. Stein, *Nature (Lond.) 257,* 764–767 (1975).

Stetler, D. A., & K. M. Rose, *Biochemistry 21,* 3721–3728 (1982).

Stiles, C. D., *Cell 33,* 653–655 (1983).

Stiles, C. D., B. H. Cochran, & C. D. Scher, in *The Cell Cycle,* P. C. L. John, ed., SEB Seminar Series, 10, pp. 119–138 (1981).

Stocum, D. L., *Dev. Biol. 79,* 276–295 (1980).

Stocum, D. L., *J. Embryol. Exp. Morphol. 71,* 193–214 (1982).

Stocum, D. L., & J. F. Fallon, *J. Embryol. Exp. Morphol. 69,* 7–36 (1982).

Streumer-Svobodova, Z., F. A. C. Wiegent, A. A. M. S. van Dongen, & R. van Wijk, *Biochimie 64,* 411–418 (1982).

Stryer, L., *Biochemistry,* 2d ed., W. H. Freeman, San Francisco, California (1981).

Stumph, W. E., M. Baez, W. G. Beattie, M. J. Tsai, & B. W. O'Malley, *Biochemistry 22,* 306–315 (1983).

Sulston, J. E., & J. G. White, *Dev. Biol. 78,* 577–597 (1980).

Sussman, M., in *Development of* Dictyostelium discoideum, W. F. Loomis, ed., Academic Press, New York, pp. 353–385 (1982).

Tabor, C. W., & H. Tabor, *Annu. Rev. Biochem. 45,* 285–306 (1976).

Tada, T., & K. Okamura, *Adv. Immunol. 28,* 1–87 (1979).

Tait, J. F., S. A. S. Tait, & J. B. G. Bell, *Essays in Biochem. 16,* 99–174 (1980).

Takaoka, K., H. Yoshikawa, N. Shimizu, K. Ono, K. Amitani, Y. Nakata, & Y. Sakamoto, *Biomed. Res. 2*, 466–471 (1981).

Tamiya, H., *Symp. Soc. Exp. Biol. 17*, 188–214 (1963).

Tanaka, K., L. Waxman, & A. L. Goldberg, *J. Cell Biol. 96*, 1580–1585 (1983).

Taniguchi, T., H. Matsui, T. Fujita, C. Takaoka, N. Kashima, R. Yoshimoto, & J. Hamuro, *Nature (Lond.) 302*, 305–310 (1983).

Tank, P. W., & N. Holder, *Q. Rev. Biol. 56*, 113–142 (1981).

Tassava, R. A., & C. L. Olsen, *Differentiation 22*, 151–155 (1982).

Tata, J. R., *Biochem. Aspects Horm. Actions 1*, 89–133 (1970).

Tata, J. R., *Biol. Rev. 55*, 285–319 (1980).

Tata, J. R., *Nature (Lond.) 298*, 707–708 (1982).

Tauber, R., & W. Reutter, *Eur. J. Biochem. 83*, 37–45 (1978).

Tedeschi, M. V., D. A. Colbert, & N. Fausto, *Biochim. Biophys. Acta 521*, 641–649 (1978).

Teng, N. N. H., & L. B. Chen, *Proc. Natl. Acad. Sci. 72*, 413–417 (1975).

Terasima, T., & L. J. Tolmach, *Nature (Lond.) 190*, 1210–1211 (1961).

Thelander, L., & P. Reichard, *Annu. Rev. Biochem. 48*, 133–158 (1979).

Thom, R., *Structural Stability and Morphogenesis*, W. A. Benjamin, Reading, Mass. (1975).

Thoma, F., Th. Koller, & A. Klug, *J. Cell Biol. 83*, 403–427 (1979).

Thomas, G., J. Martin-Pérez, M. Siegmann, & A. Otto, *Cell 30*, 235–242 (1982).

Thomas, J. O., & R. J. Thompson, *Cell 10*, 633–640 (1977).

Thomopoulos, P., F. C. Kosmakos, I. Pastan, & E. Lovelace, *Biochem. Biophys. Res. Commun. 75*, 246–252 (1977).

Thompson, d'Arcy W., *On Growth and Form*, abr., J. T. Bonner, ed., Cambridge University Press, Cambridge, England (1961).

Thornburg, W., & T. J. Lindell, *J. Biol. Chem. 252*, 6660–6665 (1977).

Thornton, C. S., *J. Exp. Zool. 134*, 357–381 (1957).

Thorssen, A., & R. L. Hintz, *N. Eng. J. Med. 297*, 908–912 (1977).

Thrower, S., & M. G. Ord, *Biochem. J. 144*, 361–369 (1974).

Thrower, S., & M. G. Ord, *Biochem. Soc. Trans. 3*, 724–727 (1975).

Thrower, S., M. G. Ord, & L. A. Stocken, *Biochem. Pharmacol. 22*, 95–100 (1973).

Thuiller, L., F. Garreau, M. Hamet, & P. Cartier, *Exp. Cell Res. 141*, 341–349 (1982).

Tickle, C., *Am. Sci. 69*, 639–646 (1981).

Tickle, C., D. Summerbell, & L. Wolpert, *Nature (Lond.) 254*, 199–202 (1975).

Tobin, A. J., *Dev. Biol. 68*, 47–58 (1979).

Tosi, M., R. A. Young, O. Hagenbüchle, & V. Schibler, *Nucleic Acids. Res. 9*, 2313–2323 (1981).

Trainin, N., M. Small, & A. I. Kook, in *B and T Cells in Immune Recognition*, F. Loor & G. E. Roelants, eds., Wiley, New York, pp. 83–102 (1977).

Traub, O., P. M. Drüge, & K. Willecke, *Proc. Natl. Acad. Sci. 80*, 755–759 (1983).

Trembley, A. *Mémoires pour servir á l'histoire naturelle d'une genre de polypes d'eau douce, á bras en forme de cornes*, J & H Verbeck, Leyden (1744).

Trendelenburg, M. F., & R. G. McKinnell, *Differentiation 15*, 73–95 (1979).

Trentin, J. J., *Am. J. Path. 65*, 621–628 (1971).

Trevillyan, J. M., & C. V. Byus, *Biochim. Biophys. Acta. 762*, 187–197 (1983).

Truman, D. E. S., *The Biochemistry of Cytodifferentiation*, Blackwell Scientific Publications, Boston (1974).

Tsanev, R., in *Cell Cycle and Cell Differentiation*, J. Reinert & H. Holtzer, eds., Springer-Verlag, New York, pp. 197–248 (1975).

Tsien, R. Y., T. Pozzan, & T. J. Rink, *Nature (Lond.) 295*, 68–71 (1982).

Turing, A. M., *Phil. Trans. (Roy) 237B*, 37–52 (1952).

Tyrsted, G., & B. Munch-Petersen, *Nucleic Acids Res. 4*, 2713–2723 (1977).

Unanue, E. R., *Adv. Immunol. 31*, 1–136 (1981).

Ungewickell, E., & D. Branton, *Nature (Lond.) 289*, 420–422 (1981).

Unwin, P. N. T., & G. Zampighi, *Nature (Lond.) 283*, 545–549 (1980).

Varon, S. S., & R. P. Bunge, *Annu. Rev. Neuro. Sci. 1*, 327–361 (1978).

Vaughan, M., & J. Moss, *Curr. Top. Cell Regul. 20*, 205–246 (1981).

Vedel, M., M. Gomez-Garcia, M. Sala, & J. M. S. Trepat, *Nucleic Acids Res. 11*, 4335–4354 (1983).

Verly, W. G., Y. Deschamps, J. Pushpathadam, & M. Desrosiers, *Can. J. Biochem. 49*, 1376–1383 (1971).

Viceps-Madora, D., K.-Y. Chen, H. R. Tsou, & E. S. Canellakis, *Biochim. Biophys. Acta 717*, 305–315 (1982).

Virolainen, M., *Exp. Cell Res. 33*, 588–591 (1964).

Virtanen, I., T. Vartio, R. A. Badley, & V. P. Lehto, *Nature (Lond.) 298*, 660–663 (1982).

Vogelstein, B., & B. F. Hunt, *Biochem. Biophys. Res. Commun. 105*, 1224–1232 (1982).

Vogelstein, B., D. M. Pardoll, & D. S. Coffey, *Cell 22*, 79–85 (1980).

Wallace, H., *Vertebrate Limb Regeneration*, Wiley, New York (1981).

Walmsley, R. M., J. W. Szostak, & T. D. Petes, *Nature (Lond.) 302*, 84–86, (1983).

Walton, G. M., & G. N. Gill, *J. Biol. Chem. 258*, 4440–4446 (1983).

Ward, P. A., I. H. Lepow, & L. J. Newman, *Am. J. Pathol. 52*, 725–736 (1968).

Warner, A. E., & P. A. Lawrence, *Cell 28*, 243–252 (1982).

Watanabe, H., D. N. Orth, & D. O. Toft, *J. Biol. Chem. 249*, 7625–7630 (1973).

Waterborg, J. M., & H. R. Matthews, *Biochemistry 22*, 1489–1496 (1983).

Waterborg, J. H., & H. R. Matthews, *Exp. Cell Res. 138*, 462–466 (1982).

Waterlow, J. C., & A. A. Jackson, *Br. Med. Bull. 37*, 5–10 (1981).

Weatherall, D. J., & J. B. Clegg, *Thalassemia Syndromes*, 3d ed., Blackwell Scientific Publications, Boston (1981).

Weatherbee, J. A., *Int. Rev. Cytol. Suppl. 12*, 113–176 (1981).

Webb-Walker, B., L. Lothstein, C. L. Baker & W. M. Le Stourgeon, *Nucleic Acids Res. 8*, 3639–3657 (1980).

Weeds, A., *Nature (Lond.) 296*, 811–816 (1982).

Weintraub, H., *Nucleic Acids Res. 8*, 4745–4753 (1980).

Weintraub, H., A. Larsen, & M. Groudine, *Cell 24*, 333–344 (1981).

Weisbrod, S., *Nucleic Acids Res. 10*, 2017–2042; *Nature (Lond.) 297*, 289–295 (1982).

Weisbrod, S., M. Groudine, & H. Weintraub, *Cell 19*, 289–301 (1980).

Weisbrod, S., & H. Weintraub, *Cell 23*, 391–400 (1981).

Weiss, P. A., *Differentiation 1*, 3–10 (1973).

Weiss, P., & J. L. Kavenau, *J. Gen. Physiol. 41*, 1–47 (1957).

Weller, M., *Protein Phosphorylation*, Pion Press, London (1979).

Wessells, N. K., *Tissue Interactions and Development*, W. A. Benjamin, Reading, Mass. (1977).

Wetmur, J. G., *Annu. Rev. Biophys. Bioeng. 5*, 337–361 (1976).

Wheals, A. E., *Mol. Cell Biol. 2*, 361–368 (1982).

Wheals, A., & B. Silverman, *J. Theor. Biol. 97*, 505–510 (1982).

White, A., in *Biochemical Actions of Hormones, VIII*, G. Litwak, ed., Academic Press, New York, pp. 1–46 (1981).

Whitfield, J. F., A. L. Boynton, J. P. MacManus, R. H. Rixon, M. Sikorska, B. Tsang, & P. R. Walker, *Ann. N.Y. Acad. Sci. 339*, 216–240 (1980).

Widmann, J. J., & H. D. Fahimi, in *Liver Regeneration after Experimental Injury*, R. Lesch & W. Reutter, eds., Stratton, New York pp. 89–98 (1975).

Wiebkin, O. W., T. E. Hardingham, & H. Muir, *Extracellular Matrix Influences on Gene Expression*, H. C. Slavkin & R. C. Greulich, eds., Academic Press, New York, pp. 209–223 (1975).

Wigglesworth, V. B., *J. Exp. Biol. 14*, 364–381 (1937).

Wigler, M. H., *Cell 24*, 285–286 (1981).

Wijk, van R., G. Zoutewelle, N. Defer, L. Tichonicky, & J. Kruh, *Biochimie 61*, 711–717 (1979).

Wilcox, M., D. L. Brower, & R. J. Smith, *Cell 25*, 159–164 (1981).

Wilkes, P. R., & G. D. Birnie, *Nucleic Acids Res. 9*, 2021–2035 (1981).

Wilkes, P. R., G. D. Birnie, & R. W. Old, *Exp. Cell Res, 115*, 441–444 (1978).

Williamson, J. R., R. H. Cooper, & J. B. Hoek, *Biochim. Biophys. Acta 639*, 243–295 (1981).

Willingham, M. C., & I. Pastan, *Cell 21*, 67–77 (1980).

Witschi, H. P., *Biochem. J. 120*, 623–634 (1970).

Wittig, B., & S. Wittig, *Cell 18*, 1173–1183 (1979).

Wittig, B., & S. Wittig, *Nature (Lond.) 297*, 31–38 (1982).

Wold, R., *Annu. Rev. Biochem. 50*, 783–814 (1981).

Wolpert, L., *Curr. Top. Dev. Biol. 6*, 183–224 (1971).

Wolpert, L., *J. Theor. Biol. 25*, 1–47 (1969).

Wolpert, L., J. Lewis & D. Summerbell, *CIBA Symposium 29 (new series)*, 95–119 (1975).

Wool, I. G., *Annu. Rev. Biochem. 48*, 719–754 (1979).

Wu, C., Y.-C. Wong, & S. C. R. Elgin, *Cell 16*, 807–814 (1979).

Wu, R. S., & W. M. Bonner, *Cell 27*, 321–330 (1981).

Wu, R. S., S. Tsai, & W. M. Bonner, *Cell 31*, 367–374 (1982).

Yamada, K. M., S. K. Akiyami, & H. Hayashi, *Biochem. Soc. Trans. 9*, 506–518 (1981).

Yanishevsky, R. M., & G. H. Stein, *Int. Rev. Cytol. 69*, 223–259 (1981).

Yau, P., B. S. Imai, A. W. Thorne, G. H. Goodwin, & E. M. Bradbury, *Nucleic Acids Res. 11*, 2651–2664 (1983).

Yee, A. G., & J. B. Revel, *J. Cell Biol. 78*, 554–564 (1978).

Yen, A., & A. B. Pardee, *Science 204*, 1315–1317 (1979).

Yukioka, M., T. Hatayama, & A. Inoue, *J. Mol. Biol. 155*, 429–446 (1982).

Yutani, Y., Y. Tei, M. Yukioka, & A. Inoue, *Arch. Biochem. Biophys. 218*, 409–420 (1982).

Zachau, H. G., & T. Igo-Kemenes, *Cell 24*, 597–598 (1981).

Zalin, R., *Dev. Biol. 71*, 274–288 (1979).

Zehner, Z. E., & B. M. Paterson, *Proc. Natl. Acad. Sci. 80*, 911–915 (1983).

Zieve, G. W., *Cell 25*, 296–297 (1981).

Zieve, G., & S. Penman, *J. Mol. Biol. 145*, 501–523 (1981).

Zigmond, S. H., *J. Cell Biol. 75*, 606–616 (1977).

Zigmond, S. H., in *Cell Behavior*, R. Bellairs, A. Curtis, & G. Dunn, eds., Cambridge University Press, New York, pp. 183–202 (1982).

Ziie, M. H., & M. E. Cullum, *Proc. Soc. Exp. Biol. Med. 172*, 139–152 (1983).

Zlatanova, J. S., *FEBS Lett. 112*, 199–202 (1980).

Zongza, V., & A. P. Mathias, *Biochem. J. 179*, 291–298 (1979).

Zwilling, E., *Biol. Bull. 124*, 368–378 (1939)

GLOSSARY

AGONIST.   Opposite of antagonist. Usually applied to neuotransmitters binding to a membrane site and eliciting stimulatory response.

AUTOPHAGY.   Self-digestion of cell contents, usually involving lysosomes.

BASEMENT MEMBRANE.   Membrane adjacent to and produced in part by, epithelial cell layer.

CHEMOTAXIS.   Movement of cells (usually up a concentration gradient) due to remote cellular release of an attractant.

CLEAVAGE CYCLE.   Early cell division cycles following fertilization. Cycles are characterized by synchronous division of cells, without significant increase in cell mass.

CYTOKINESIS.   Process leading to the separation of two daughter cells following mitosis.

ECTODERM, ENDODERM.   Cells in or derived from outer or inner layer, respectively, of cell mass during embryonic development.

ENDOCYTOSIS, EXOCYTOSIS.   Process of vesicle formation from plasma membrane causing engulfed material to enter or leave (respectively) the cell.

ENDONUCLEASES, EXONUCLEASES.   Hydrolases attacking internucleotide bonds within DNA (or RNA). Exonucleases attack from 5' or 3' ends of nucleic acids.

EXON.   Nucleotide sequence which is retained in mRNA after processing and from which protein is translated.

EXTRACELLULAR MATRIX.   Material adherent to and released from cells. Contains collagen, glycopeptides, and various glycosaminoglycans.

GENE AMPLIFICATION.   A process that results in multiple copies of genes being incorporated into the genome, for example, Amphibian genes for rRNA during Xenopus oocyte development.

HELPER T CELLS.   Class of T lymphocytes which play an essential part in reception and stimulation of antibody production by B lymphocytes.

HILL COEFFICIENT.   Kinetic parameter indicating extent of cooperativeness of ligand binding in allosterically regulated proteins.

HOMOKARYONS, HETEROKARYONS.   Cells resulting from fusion of homologous or heterologous (respectively) cells.

HYPERPLASIA.   Increase in tissue mass associated with increase in cell number.

HYPERTROPHY.   Increase in tissue mass associated with increase in cell size.

IMAGINAL DISC.   Patches of cells in insect larvae from which organs develop.

INTRON.   Nucleotide sequence within a gene which is transcribed but subsequently eliminated from pre-mRNA by splicing out.

IONOPHORE.   Substance facilitating plasma membrane translocation of a solute.

LAMELLIPODIA.   Platelike protrusion of the cell membrane observed by electron microscopy.

LECTIN.   Multivalent carbohydrate binding protein which agglutinates cells with complementary glycosylated residues in their plasma membranes.

MITOGEN.   Substance inducing mitosis.

MITOTIC SHAKING.   A means of selecting synchronous cells from monolayer cultures because of the reduced adherence of (rounded up) cells in mitosis.

MONOCLONAL ANTIBODY.   Antibody produced by clone of cells picked from cell cultures derived from fusing spleen cells from sensitized animal with myeloma cell line.

MORPHOGEN.   Agent involved in embryonic development.

PERIOSTEUM.   Fibrous membrane covering the bone surface.

PLASMODIUM.   A coalescence of cells forming a multinucleate organism with a single surrounding membrane.

PLEIOTROPY.   Multiple diverse consequences of a single biological event.

POIKILOTHERM.   Animal whose temperature varies with that of the surrounding medium.

POLYTENE CHROMOSOME.   Giant chromosomes in some somatic cells of Diptera, for example, salivary gland.

PROPHASE, METAPHASE, ANAPHASE, AND TELOPHASE.   Successive stages in mitosis.

PURINERGIC NERVES.   Nerves which release purine derivatives on stimulation.

SCHWANN CELLS.   Cells that surround medulated nerve fibers and synthesize the myelin sheath.

SYNERGISM.   Interaction of two or more substances to give responses greater than the sum of the separate effects.

INDEX